CRITICAL THINKING & LOGICAL REASONING WORKBOOK-5

5

GIFT OF LOGIC™ SERIES

An Essential Resource for Everyone

Boost Your Thinking Skills

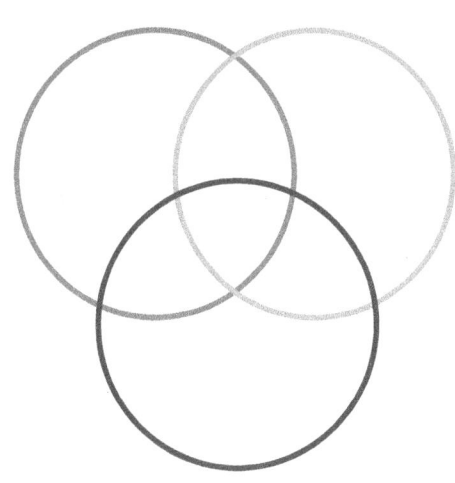

Verbal Reasoning

Analytical Reasoning

Pictorial Reasoning

THIRD EDITION

| FOR GRADES 3-5 | STUDENTS, TEACHERS, AND PARENTS |

Ranga Raghuram

 GIFT OF LOGIC™

Gift Of Logic, Inc

http://www.giftoflogic.com
sales@giftoflogic.com

Critical Thinking and Logical Reasoning Workbook-5
ISBN-13: 978-1494832322
ISBN-10: 1494832321

Third Edition
1-2014

Copyright © 2009 Gift Of Logic, Inc. All rights reserved. No part of this publication may be reproduced, stored in a retrieval system, transmitted in any form or by any means, electronic, mechanical, photocopying, recording or otherwise, without the written permission of the publisher.

License: This book is licensed for use by one person only. Use of this book in a group setting (classroom, workshop, etc) without the written permission of the publisher is prohibited. Unauthorized duplication is strictly prohibited by law. Contact the publisher at sales@giftoflogic.com for classroom/school/group licensing.

GIFT OF LOGIC™
CRITICAL THINKING & LOGICAL REASONING CURRICULUM
12 WORKBOOKS TO BOOST YOUR THINKING SKILLS

For Kindergarten, Grade 1, and Grade 2

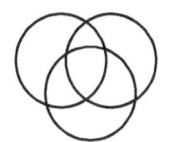

Workbook# 0

Verbal Reasoning	Finding the truth, Inferencing, Analogies, Synonyms and Antonyms, Agree/Disagree
Analytic Reasoning	Memory drill, Decision making, Positioning, Sudoku
Pictorial Reasoning	Connect the dots, Mazes, Picture Sequence, Spot the difference, etc

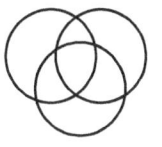

Workbook# 1

Verbal Reasoning	Finding the truth, Inferencing, Analogies, Synonyms and Antonyms, Agree/Disagree
Analytic Reasoning	Sorting, Positioning, Picking, Assorted problems, Numeric and Alphabetic Sudoku
Pictorial Reasoning	Picture Sequence, Spot the difference, Odd picture

Workbook# 2

Verbal Reasoning	Finding the truth, Classification, Direct and Inverse relationship, Inferencing, Analogies, Agree/Disagree
Analytic Reasoning	Sequencing, Scheduling, Strategy, Picking, etc
Pictorial Reasoning	Picture Analogy, Odd picture, Pattern matching, etc

For Grade 3, Grade 4, and Grade 5

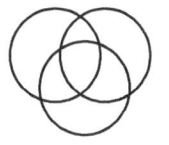

Workbook# 3

Verbal Reasoning	Not, And, Or, If .. then, Conditional inferencing, Unconditional inferencing, Symbolic Logic
Analytic Reasoning	Lists, Sequencing, Grouping, Venn Diagrams, Graph logic, Number logic, Letter logic, Sudoku
Pictorial Reasoning	Picture sequence, Picture analogy, Odd picture, Picture difference, Pattern matching

Workbook# 4

Verbal Reasoning	Contradiction, Converse, Inverse, Contrapositive, Conditional inferencing, Symbolic Logic
Analytic Reasoning	Scheduling, Looping, FIFO, LIFO, Correlation, Venn Diagram, Graph logic, Number logic, Sudoku, etc
Pictorial Reasoning	Picture sequence, Picture analogy, Odd picture, Picture difference, Pattern matching

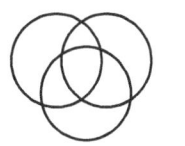

Workbook# 5

Verbal Reasoning	Biconditional, Categorical inferencing, Cause and Effect, Symbolic Logic, Agree/Disagree, Word and Sentence analogy
Analytic Reasoning	Correlation, Grouping, Venn Diagrams, Graph logic, Number logic, Letter logic, Sudoku, etc
Pictorial Reasoning	Picture sequence, Picture analogy, Odd picture, Picture difference, Pattern matching

********* Essential resource for everyone *********
*http://www.giftoflogic.com *sales@giftoflogic.com

GIFT OF LOGIC™
CRITICAL THINKING & LOGICAL REASONING CURRICULUM
12 WORKBOOKS TO BOOST YOUR THINKING SKILLS

For Grades 6-12, College/University Students, Adults

Primer

Prereq

Verbal Reasoning	Logical Operators, Conditional, Categorical and Causal reasoning, Validity, Fallacies, Symbolic Logic
Analytic Reasoning	Positioning, Grouping, Sudoku
Pictorial Reasoning	Pattern perception, Figure formation, Paper folding and cutting, Figure matrix, Rule detection

Workbook# 6

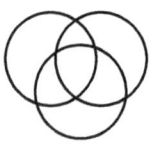

Verbal Reasoning	Arguments-Main point, Must be true, Cannot be true
Analytic Reasoning	Positioning, Grouping, Sudoku
Pictorial Reasoning	Pattern perception, Figure formation, Paper folding and cutting, Figure matrix, Rule detection

Workbook# 7

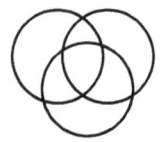

Verbal Reasoning	Arguments-Strengthening, Weakening
Analytic Reasoning	Positioning, Grouping, Sudoku
Pictorial Reasoning	Pattern perception, Figure formation, Paper folding and cutting, Figure matrix, Rule detection

Workbook# 8

Verbal Reasoning	Arguments - Controversy, Paradox
Analytic Reasoning	Positioning, Grouping, Sudoku
Pictorial Reasoning	Pattern perception, Figure formation, Paper folding and cutting, Figure matrix, Rule detection

Workbook# 9

Verbal Reasoning	Arguments- Assumptions, Reasoning strategy
Analytic Reasoning	Positioning, Grouping, Sudoku
Pictorial Reasoning	Pattern perception, Figure formation, Paper folding and cutting, Figure matrix, Rule detection

Workbook# 10

Verbal Reasoning	Arguments-Flawed reasoning, Analogous reasoning
Analytic Reasoning	Positioning, Grouping, Sudoku
Pictorial Reasoning	Pattern perception, Figure formation, Paper folding and cutting, Figure matrix, Rule detection

********* Essential resource for everyone *********
Get the GIFT OF LOGIC™ today !
*http://www.giftoflogic.com   *sales@giftoflogic.com

© Gift Of Logic, Inc * Copying prohibited

Dear Reader:

Your decision to purchase this book is commendable. You now have in your hands, a comprehensive, easy-to-read book in Critical thinking and Logical reasoning that will introduce you to three different areas of thinking and reasoning - Verbal, Analytic and Pictorial. Solving problems in Verbal Reasoning is important to develop a critical mind. Solving problems in Analytic Reasoning is important to develop a flexible and resourceful mind. Solving problems in Pictorial Reasoning is important to develop a visually alert mind.

This book is presented in a workbook format to help you progress quickly. Parents and teachers are urged to complete the exercises ahead of the student and assist them whenever necessary with the help of detailed answers provided at the end of the book. This book can be used as a supplementary resource in the regular class room or it can be used during winter and summer vacations. College/University students, working professionals and retired individuals will also find this book very useful in enhancing their problem solving abilities, confidence and general intellect.

Critical thinking and Logical reasoning must be practiced consistently to develop strong cognitive skills. After completing the exercises in this book, continue to read the other books in this series to get familiar with different types of Logical reasoning problems.

This book is one in a series of twelve books. Please visit the website at http://www.giftoflogic.com to get more information on these books, to understand your benefits, and to provide feedback.

Happy thinking and reasoning !

TABLE OF CONTENTS

Verbal Reasoning

Conditional statements (Only If)..9

Biconditional statements (If and only if)..10

Inferencing - Conditional..11

Categorical statements..12
    All
    Some
    None

Inferencing-Categorical..17

Causal statements
    Cause and effect...21
    Finding the root cause..25

Inferencing-Causal..26

Agree or Disagree...28

Sentence Analogy...30

Word Analogy..32

TABLE OF CONTENTS

Analytic Reasoning

List Processing...35
Sequencing...40
Correlation..42
Grouping..45
Venn Diagram...46
Graph Logic..49
Number Logic...56
Letter Logic..59
Sudoku...62

Pictorial Reasoning

Picture Sequence..69
Picture Analogy..72
Odd Picture...75
Picture Difference...77
Pattern Matching...80

Answers

Verbal...82
Analytic..95
Pictorial..125

Certification of Completion

Name ——————————————— Date ———————————————

VERBAL REASONING

Name _____   Date_____

CONDITIONAL STATEMENTS - Only If

Conditional statements that are expressed using "only if" have a special meaning. "Only if" can appear in the beginning or middle of a statement as shown below.

> P only if Q.
> Only if P then Q.

The statement "P only if Q" means the following:
  P cannot happen alone.
  If P happens then Q happens
  Q can happen alone.

This is the same as the conditional "If P then Q". (see workbook# 3 and 4) So, "P only if Q" is logically the same as "if P then Q". Note the positions of P and Q in the "only if" statement and the "if.. then statement". They are switched.

"Only if" in the middle: P only if Q
Conversion to if..then: If P then Q (P → Q)

"Only if" in the beginning: Only if P then Q
Conversion to if..then: If Q then P (Q → P)

Example:
Only if: Rina will play the game only if Sam plays.
If..then: If Rina plays then Sam will play.
Symbolic: Rina → Sam
Contrapositive: If Sam does not play, then Rina does not play.
Symbolic: ~ Sam → ~Rina

Name _____   Date_____

BICONDITIONAL STATEMENTS - If and only if (↔)

Conditional statements that are expressed using "If and only if" are called biconditionals. The symbol for biconditional is ↔.

"If and only if P then Q" can be broken down into two statements:
    1) <u>If</u> P then Q
         <u>and</u>
    2) <u>Only if</u> P then Q

The second statement can be converted to "if..then" as "if Q then P".

If and only if P then Q is the same as
    if P then Q (P → Q)
      and
    if Q then P (Q → P)

"If and only if P then Q" is represented by the biconditional symbol ↔ as follows: P ↔ Q

From above discussion, P ↔ Q is the same as P → Q & Q → P. Contrapositive inferences of these conditionals are valid.

<u>Example:</u>
If and only if: If and only if there is light, we can see.
Symbolic: light ↔ see
If..then: if there is light we can see; if we see then there is light
If..then symbolic: light → see ; see → light
Contrapositive: ~see → ~light (if we can't see then there is no light)
        ~light → ~see (if there is no light, then we can't see)

Verbal Reasoning

Name _____   Date_____

INFERENCING - Only If, If and only if

1

Trash will be picked up only if it is a Friday.

If the above information is true, which one of the following must be true?

A) If it is a Friday, then trash will be picked up.
B) If it is not a Friday, then trash will not be picked up.
C) If trash is not picked up, then it is not a Friday.

2

Sonya will sing if and only if Tonya sings.

If the above information is true, which one of the following must be true?

A) If Tonya does not sing, then Sonya will sing
B) If Sonya does not sing, then Tonya will sing.
C) If Tonya sings, then Sonya will sing.

Name ———————————————   Date————————————

CATEGORICAL STATEMENTS

Categorical statements refer to one or more categories (also called groups or sets).

The relationship between two categories P and Q can be represented as follows:

>All P are Q.
>Some P are Q.
>No P is Q.

We will use Venn diagrams to represent the relationships between the different categories, and to draw inferences.

Examples:
>All monkeys are animals. (categories: monkeys and animals)
>Some animals are herbivores. (categories: animals and herbivores)
>No car is an airplane. (categories: cars and airplanes)

Make up your own categorical statements of the type shown:

All P are Q:

Some P are Q:

No P is Q:

Verbal Reasoning

Name ——————————————— Date———————————

CATEGORICAL STATEMENTS   All P are Q

When all members of category P belong to category Q, the categorical statement "All P are Q" describes their relationship. This can be represented using a Venn as follows:

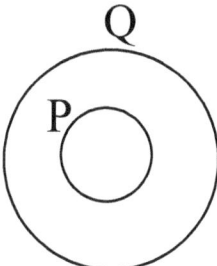

The two circles P and Q represent categories P and Q respectively. Note that circle P is completely inside circle Q because "All P are Q".

Inferencing from categorical statements is mainly done with the help of Venn diagrams and in some cases using conditional statements.

The converse of categorical "All P are Q" is an invalid inference.
    Categorical: All monkeys are animals
    Converse: All animals are monkeys.

This is an invalid inference, since we know that a Lion is not a monkey. We can also infer this from the Venn diagram. The "x" represents an "animal" and it is clear that it is not a monkey, since it is outside the circle represented by monkeys.

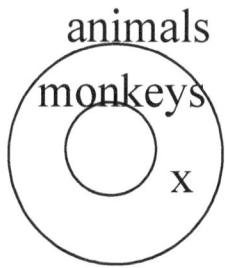

Verbal Reasoning
© Gift Of Logic, Inc * Copying prohibited

CATEGORICAL STATEMENTS   All P are Q

"All P are Q" can be converted to a conditional statement "If it is a P then it must be a Q".

"All P are Q" is logically the same as "If P then Q"

Categorical: All P are Q
Conditional: If P then Q (P → Q)

Categorical: All P are Q
Inference: If it is not a Q, then it is not a P. Is this a valid inference?

In the Venn diagram, "x" represents an item that is not Q. Since "All P are Q", this "x" cannot be inside P. So, we can infer that "if it is not a Q, then it is not a P"

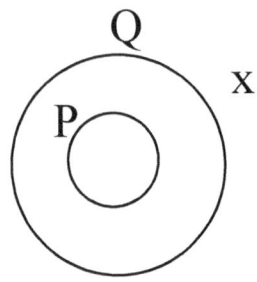

A quicker way to verify the inference is as follows.
   All P are Q: P → Q
   Contrapositive: ~Q → ~P

This contrapositive can be stated as "If it is not a Q, then it is not a P". Therefore, this is a valid inference.

Verbal Reasoning

Name ———————————————   Date ———————————

CATEGORICAL STATEMENTS   Some P are Q

When some members of category P belong to category Q, the categorical statement "Some P are Q" describes their relationship. This is represented by two overlapping circles with a common area between them and an "x" to identify one member common to both categories P and Q.

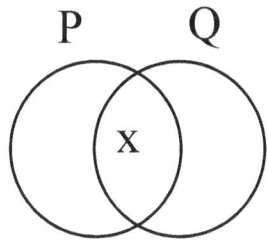

The converse of categorical " Some P are Q" is a valid inference.
Categorical: Some P are Q
Converse: Some Q are P
The converse is valid as clearly seen from the Venn diagram. If some P are Q, then some Q are P.

Categorical: Some P are Q
Inference: All P are Q
Is the above inference correct?

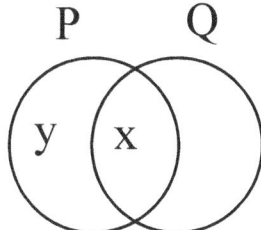

The inference is incorrect. There can be members like "y" that are P but not Q. That is, there may be some P that are not Q.

Verbal Reasoning
© Gift Of Logic, Inc * Copying prohibited

Name _____   Date_____

CATEGORICAL STATEMENTS   No P is Q

When two categories P and Q have nothing in common between them, their relationship is described by the categorical statement "No P is Q". It is represented using a Venn diagram by two circles that do not have anything in common between them.

The converse of categorical statement "No P is Q" is a valid inference.

Categorical: No P is Q.
Converse: No Q is P.

The converse is a valid inference as can be seen from the Venn diagram. Since the circles representing the categories have nothing in common, none of P is Q also means that none of Q is P.

The categorical statement "No P is Q" can be converted to the following conditional statements:

Categorical: No P is Q
If then: If it is a P then it is not a Q.   P → ~Q
Contrapositive: If it is a Q, then it is not a P.   Q → ~P

Verbal Reasoning
© Gift Of Logic, Inc * Copying prohibited

Name ——————————————   Date ——————————————

INFERENCING - Categorical

Read the following categorical statements and answer if the statements shown are valid or invalid inferences. Draw Venn diagrams to verify the inferences.

1

All Italians are Europeans.

1) If you are an Italian, you are an European.
   A) Valid B) Invalid

2) If someone is an European, that person is an Italian.
   A) Valid B) Invalid

2

No criminal is innocent.

1) There are some innocent people who are criminals.
   A) Valid B) Invalid

2) There is no innocent person who is a criminal.
   A) Valid B) Invalid

3

Some clowns are students.

1) Some students are clowns.
   A) Valid B) Invalid

2) All students are clowns.
   A) Valid B) Invalid

Verbal Reasoning   Answers-83
© Gift Of Logic, Inc * Copying prohibited

Name _____   Date _____

INFERENCING - Categorical

Read the following categorical statements and answer if the statements shown are valid or invalid inferences. Draw Venn diagrams to verify the inferences.

4

Some rats have long tails.

1) Some animals that have long tails are rats.
    A) Valid   B) Invalid

2) All animals that have long tails are rats.
    A) Valid   B) Invalid

5

All politicians are human beings.

1) If there is a human being, he must be a politician.
    A) Valid   B) Invalid

2) There is no human being who is also a politician.
    A) Valid   B) Invalid

Name ——————————————   Date——————————

INFERENCING - Categorical

6 Question 6-10 involve the usage of All, Some and None.

All the locks can be opened with this key. Bobby has a lock.

If the above statements are true, which of the following <u>cannot be true?</u>
    A) Bobby's lock can be opened with this key.
    B) Bobby's lock cannot be opened with this key.

7

All the boys must stand behind the girls. Chad is a boy.

If the above statements are true, which of the following <u>cannot be true?</u>
    A) A girl stood in front of Chad.
    B) A girl stood behind Chad.

8

No girl can be in the sports team. Angelina is a girl. Brad is a boy.

If the above statements are true, which of the following <u>cannot be true?</u>
    A) Angelina is in the sports team.
    B) Brad is in the sports team.

INFERENCING - Categorical

9

Some reptiles are dangerous. Some are not. The ones that are not dangerous make good pets. The lizard is not a dangerous reptile.

If the above statements are true, which of the following cannot be true?
  A) Lizards make good pets.
  B) Lizards do not make good pets.

10

No student can use a cell phone in the class. Anita is a student.

If the above statements are true, then which of the following cannot be true?
  A) Anita can use her cell phone in the class.
  B) Anita cannot use her cell phone in the class.

Name _____   Date _____

CAUSAL STATEMENTS

Causal statements describe cause and effect relationship between events. A few examples of causal statements are shown below.

    Heavy rain caused the flooding.
    Loud noise caused my headache.

The following words are typically used to describe cause and effect relationships:
    caused by, because of, effect of, due to, responsible for

It is important to ascertain what the cause is and what the effect is. From a time perspective, the cause occurs first and the effect happens after the cause. The "effect" happens because of the "cause". The "cause" makes the "effect" happen. The reason for the "effect" is the "cause".

Causal statements are such that, we cannot say with one hundred percent certainty if they are true or false. Instead we can say that it is highly probable to be true or false. For example, if you say that the loud noise caused your headache, then how certain are you that nothing else caused your headache? If a doctor tells you that your headache was caused by a sinus infection, then your causal statement is not true. But, if the doctor does not find any sinus infection and you have a high level of confidence that the loud noise was the only reason for your headache, then you can make that conclusion.

Symbolic Representation: The symbol c→ can be used to represent cause and effect. If P and Q are events, then the causal statement "P caused Q" can be represented as P c→ Q. P occurs first and then Q occurs.

Verbal Reasoning

CAUSE AND EFFECT

In the following table, a few causal statements are shown with their cause and effect. Complete the rest.

Causal Statement	Cause	Effect
P c→Q	P	Q
Rain causes flooding.	Rain	Flooding
Flooding is caused by rain.	Rain	Flooding
Flooding is the effect of rain.	Rain	Flooding
The reason that we lost the game is due to poor team spirit.		
Heavy load on his shoulders is responsible for his backache.		
The effect of global warming is unpredictable weather.		
The rabbit chewed the plant causing it to die.		
Janet is having health problems because of overeating.		
Tom's carelessness caused the boat accident.		

Verbal Reasoning
© Gift Of Logic, Inc * Copying prohibited

Name _____   Date _____

INFERENCING FROM CAUSAL STATEMENTS

Converse of a causal statement cannot be inferred:
   P c→ Q
   Q c→ P is an invalid inference.

P c→ Q means P is the cause and Q is the effect. P happened first and then caused Q to happen. If this is true, then we cannot prove the converse.

Example 1:
 Causal: Defective traffic lights caused the accident. (P caused Q)
 Converse: The accident caused the lights to be defective. (Q caused P).

Example 2:
Causal: The sudden school closure caused a traffic jam in the road.
Converse: The traffic jam caused the sudden school closure.

Succession of two events does not mean that one caused the other.
If two events P and Q occur one after the other, that does not automatically imply that P caused Q. Whether P caused Q or not must be investigated thoroughly before coming to that conclusion. After thorough investigation, we may or may not come the conclusion that P caused Q.

Example: Jack fell down <u>and then</u> Jill fell down. Therefore, Jack caused Jill to fall down.

Reasoning: This is an invalid inference. There is nothing in the facts to prove that Jack caused Jill to fall down. Just because Jill fell down after Jack fell down does not mean that Jack "caused" Jill to fall down.

Verbal Reasoning
© Gift Of Logic, Inc * Copying prohibited

Name —————————— Date ——————————

INFERENCING FROM CAUSAL STATEMENTS

Correlation does not imply causation.

Just because two events are related in some way does not mean that one causes the other. That is, correlation does not imply causation.

Example: The Royal college is expensive. Graduates of the Royal college are very smart. Therefore, we can infer that the expensive education causes graduates to be very smart.

Reasoning: This inference is incorrect. The correlation between the two statements is "Royal College" and this correlation is used to infer that expensive education at Royal college causes The Royal College graduates to be very smart.

Inferences can be drawn from proven Cause-Effect relationships.

Example: Research indicates that spending money lavishly causes people to live a happy life. People living in the city of Plano spend money lavishly.

If the above facts are true, which of the following is likely to be true?
    A) Residents of Plano do not live happily.
    B) Residents of Plano live happily.

Reasoning: Note the causal relationship in the first statement. It can be represented as follows: lavish spending c→ happy life
The second statement says that residents of Plano spend money lavishly. So, we can infer that residents of Plano are likely to live a happy life.

Verbal Reasoning

FINDING THE ROOT CAUSE

When several events are interlinked in a cause-and-effect chain relationship, the root cause can be found by tracing back in the causal chain. Consider the following causal chain:

$$P \xrightarrow{c} Q \xrightarrow{c} R$$

In the above cause-effect chain, Q is the direct cause of R and P is the direct cause of Q. Also, P is the root cause of R. If the root cause P does not happen, then the effect R will not happen.

If P causes Q and Q causes R; then we can infer that P causes R.
$P \xrightarrow{c} Q$; $Q \xrightarrow{c} R$, we can infer that $P \xrightarrow{c} R$

Example:
Nails slip out constantly from the hands of construction workers onto the main road. The nails cause the car tires to go flat. Flat tires are responsible for the frustration of car owners.

$$\text{nails} \xrightarrow{c} \text{flat tires} \xrightarrow{c} \text{frustration}$$

Assume that we only know the following causal relationship:

$$\text{flat tires} \xrightarrow{c} \text{frustration}$$

So, we will naturally fix the flat tires and remove the frustration. But, the tires will keep getting flat constantly and the frustration also will occur constantly. That is why we need to investigate the root cause for the frustration by tracing back in the causal chain. When we find out that the nails dropping from the hands of the construction workers is the root cause, we can ask them to be careful not to drop the nails and thus the tires will not go flat and there will not be any frustration.

Verbal Reasoning
© Gift Of Logic, Inc * Copying prohibited

INFERENCING - Causal

In the following table, a few causal statements along with some inferences are shown. Evaluate the inferences as valid or invalid.

Causal statement	Inference	Valid/Invalid
The ball landed on the table causing the milk to spill. The milk that spilled spoiled the carpet.	The ball that landed on the table caused the carpet to be spoiled.	
A fire in the building caused the firefighters to come.	The firefighters caused the fire in the building.	
First there was an explosion, then there was smoke.	The explosion caused the smoke. The smoke caused the explosion.	
Tom pushed Harry and Harry fell on a lamp. So the lamp broke.	Tom's push caused the lamp to break.	
Pizza is tasty. Pizza is fatty.	The fat in the pizza causes the taste.	
High temperatures will cause power outage. The temperature today is very high.	There will be power outage today.	

Name _____ Date_____

INFERENCING - Causal

1

Riding bikes regularly causes leg muscles to become strong. Rosie rides her bike regularly in the park.

If the above statements are true, which of the following must be true?
    A) Rosie has strong leg muscles.
    B) Rosie does not have strong leg muscles.

2

While it is true that there was a mild earthquake when the building collapsed, experts agree that it was not the earthquake, but the poor quality of the foundation that caused its collapse.

If the above statements are true, which of the following must be true?

A) If the earthquake had not happened, the building would not have collapsed.
B) If the quality of the foundation was not poor, the building would not have collapsed.

Name _____   Date _____

AGREE-DISAGREE

In questions 1-8, read the information given and agree or disagree with it as instructed.

1 Recycling is good for the environment. So, newspapers must not be thrown into the trash bin, but instead, they must be recycled.

Agree:

2 Don't paint the wall blue. There are blue colored pictures hanging on the wall and they won't look nice on a blue wall.

Disagree:

3 Walking to school is good for our health. Moreover, by not taking our cars and motorbikes to school, we can help reduce air pollution. Therefore, we all should walk to school as much as possible.

Agree:

AGREE-DISAGREE

4 We must turn off the lights when we do not need them.

Agree:

---

5 We should maintain a good landscape in front of our house. This will make our house look beautiful.

Agree:

---

6 The ball that he threw might have hit the bulb. That is why the bulb fell down and broke into pieces.

Disagree:

---

7 I think Madrid is the capital of Egypt.

Disagree:

---

8 Doctors work very long hours. This is not good because they can make mistakes. So, doctors must not work very long hours.

Disagree:

SENTENCE ANALOGY

1 Statement: Our stomach is like a balloon.

What are being compared in the statement?

If the statement is true, then which of the following must be true?
    A) As we eat more, our stomach gets bigger.
    B) As we eat more, our stomach shrinks.

2 Statement: We grow in height the same way as trees grow in height.

What are being compared in the statement?

If the statement is true, then which of the following must be true?
    A) We keep growing tall forever.
    B) We grow tall for some time and then stop growing.

3 Statement: The purpose of our eyelids is similar to the purpose of an umbrella.

What are being compared in the statement?

If the statement is true, then which of the following must be true?
    A) Eyelids protect the eyeballs.
    B) Eyelids do not protect the eyeballs.

Verbal Reasoning   Answers-92
© Gift Of Logic, Inc * Copying prohibited

SENTENCE ANALOGY

4 The relation between Tom and Dick is similar to the relation between two adversaries.

What are being compared in the statement?

If the statement is true, then which of the following must be true?
 A) Tom and Dick work well together.
 B) Tom and Dick often quarrel with each other.

---

5 This problem is a tough nut to crack.

What are being compared in the statement?

If the statement is true, then which of the following must be true?
 A) This problem is easy to solve.
 B) This problem is difficult to solve.

---

6 Holly dresses like a fashion model.

What are being compared in the statement?

If the statement is true, then which of the following must be true?
 A) Holly's dresses are contemporary.
 B) Holly's dresses are outdated.

	WORD ANALOGY
1	Spain is to Madrid as England is to A) Dallas B) London C) Frankfurt
2	Sound is to silence as Light is to A) sun B) brightness C) darkness
3	Bird is to parrot as Animal is to A) eagle B) fox C) ostrich
4	Man is to boy as Woman is to A) child B) baby C) girl
5	Pilot is to turbulence as Sailor is to A) water B) fish C) gust
6	Plumber is to pipe as Carpenter is to A) iron B) gold C) wood
7	Lawyer is to court as Doctor is to A) hotel B) beach C) hospital
8	January is to December as Sunday is to A) Friday B) Saturday C) Monday
9	First is to last as Begin is to A) start B) middle C) end

Verbal Reasoning   Answers-94

© Gift Of Logic, Inc * Copying prohibited

WORD ANALOGY

10	introduce : recall :: insert : A) screw B) fill C) retract
11	brain : think :: stomach : A) store B) dispose C) digest
12	square : rectangle :: circle : A) triangle B) ellipse C) arc
13	eye : vision :: ear : A) noise B) hearing C) silence
14	lawyer : paralegal :: doctor : A) secretary B) nurse C) surgeon
15	virus : sickness :: earthquake : A) construction B) destruction
16	whole : part :: earth : A) universe B) sun C) continent
17	part : whole :: classrooms : A) state B) country C) school
18	book : page :: page : A) book B) page C) sentence

Name ——————————————— Date ———————————————

ANALYTICAL REASONING

Name _____   Date _____

1 LIST PROCESSING - alphabetic sorting

Sort the following list of names in ascending and descending order.

Name	Ascending	Descending
Steve		
Preety		
Bobby		
Huy		
Karan		
Arjun		
Zita		

1) Who is the first in the list when sorted in the ascending order?

2) Who is the first in the list when sorted in the descending order?

3) When sorted alphabetically in ascending order, Zita is the last in the list. So, Zita is the tallest in the group.
    A) Valid Reasoning   B) Invalid Reasoning

4) When a list that has an odd number of items is sorted in ascending order and descending order, one item in the list will have the same rank.
    A) True   B) False

Analytical Reasoning
© Gift Of Logic, Inc * Copying prohibited

Name _____ Date _____

2   LIST PROCESSING - numeric sorting

You can sort numbers in ascending order or in descending order.
Sorting from a small number to a big number is called ascending order.
Sorting from a big number to a small number is called descending order.
Sort the ID# column in ascending and descending order and write it in the respective columns.

ID#	Ascending order	Descending order
10		
7		
8		
6		
1		
4		
9		

Exits in a highway are numbered in descending order going north and in ascending order going south.

1) Harrison was driving his car north towards the mountain and the highway exits he saw were numbered 1, 2 and 3. This information
    A) must be true
    B) cannot be true

2) Julia was going south in the highway to her grandmother's home and noticed exits numbered 11, 12 and 13. This information
    A) must be true
    B) cannot be true

Name ____________________ Date____________________

3 LIST PROCESSING - mixed sorting

Name	**Age**
Steve	7
Preety	6
Bobby	5
Huy	6
Karan	7
Arjun	5
Zita	6

Sort the list shown above in ascending order by name.

Name	**Age**

Sort the list shown above in ascending order by age.

Name	**Age**

Name —————————————— Date——————————

LIST PROCESSING - mixed sorting - continued

Sort the list shown in the previous page in ascending order by age and if there is a tie, sort by name. Write the sorted names below.

Sort by Age and Name

Name	Age

Use the lists shown above and answer the following questions.

1) How many people are 6 years old?
 Which of the lists were the most useful to answer this question and why?

2) Why is the Sort by Age list different from the Sort by Age and Name list?

4 LIST PROCESSING - sorting and ranking

After sorting a list, it is sometimes necessary to rank the items in the list.

Student	Score
Cynthia	45
Lauren	35
Aparna	20
Omar	40
Josh	30
Arti	25

Sort the list in descending order based on the student scores and write their ranks.

Student	Score	Rank

Name _____   Date _____

SEQUENCING

1 The following paragraph describes a series of events that are all jumbled up. Rewrite the paragraph by arranging them in a logical sequence.

Following the Math class was the Science class. The first class in the morning was a Math class. The last class of the day was a Sports class.

---

2 Bill has an appointment with the dentist in the morning. After seeing the dentist, he will go and get groceries. After that, but before going home, he will buy a newspaper.

Which of the following indicates the correct chain of stops planned by Bill?
    A) Dentist-Groceries-Home-Newspaper
    B) Dentist-Groceries-Newspaper-Home

---

3 Jill went to the park yesterday. Day before yesterday she went to a movie. Tomorrow, she will go to the Zoo. Day after tomorrow she will go to see her grandma. But, today she is at home doing her homework.

Which of the following describes the sequence of places where Jill would be present?
    A) Park,Movie,Zoo,Grandma,Home
    B) Movie,Park,Home,Zoo,Grandma

Name —————————————— Date————————————

SEQUENCING

4 Sequence based on time

The space shuttle was moved to the launching pad at 8:00 AM. At 9:00 AM it was fuelled. At 10:00 AM the astronauts got seated. At 11:00 AM it blasted off into the blue skies.

Which of the following indicates the times at which events happened until lift off?

    A) 11:00 AM - 10:00 AM - 9:00 AM - 8:00 AM
    B) 8:00 AM - 9:00 AM - 10:00 AM - 11:00 AM

5 Sequence based on day

Jack has a fitness class every other day starting from Monday.
1) Which one of the following represents the sequence of days in which Jack has a fitness class?

    A) Tuesday, Thursday
    B) Monday, Wednesday, Friday

6 Sequence based on recurring events

Pizza is served for lunch every other day starting from Monday. Sandwich is served for lunch every Tuesday and Thursday.

Which one of the following represents the sequence of items served during the week starting from Monday?

    A) Pizza,Sandwich, Pizza,Sandwich,Sandwich
    B) Pizza, Sandwich,Pizza, Sandwich,Pizza

Analytical Reasoning
© Gift Of Logic, Inc * Copying prohibited

1 CORRELATION

Correlate the information in the tables and answer the following questions.

ID	Name
1	Jeff
2	Chen
3	Prem
4	Gonzales

ID	City	Country
1	Dallas	USA
2	Shanghai	China
3	Chennai	India
4	Madrid	Spain

1) In which city does Prem live?
2) Who lives in Dallas?
3) What is the ID of the person living in Madrid?
4) Who lives in China?

Place all the information in one table shown below.

ID	Name	City	Country
1			
2			
3			
4			

Name _____   Date_____

2 CORRELATION

Correlate the information in the following two tables and answer the questions below.

Time	Class
8:00 AM	Craft
10:00 AM	Drawing
2:00 PM	Painting
4:00 PM	Puppetry

Time	Teacher	Room
8:00 AM	Osborne	101
10:00 AM	Gupta	144
2:00 PM	Sweeney	135
4:00 PM	Gilbert	128

1) Name the teachers who will teach in the mornings.

2) Name the teachers who will teach in the afternoons.

3) In which rooms will morning classes be held?

4) In which rooms will the afternoon classes be held?

5) Name the teacher who will teach the Painting class.

6) What class will Mr. Gilbert teach in room# 128?

Analytical Reasoning
© Gift Of Logic, Inc * Copying prohibited

3 CORRELATION

Robert's class schedule for the week is shown in the following tables. Robert wears a watch to school everyday. Correlate the information in the two tables and answer the questions below.

Time	Class
8 AM	Reading
9 AM	Math
10 AM	Sports
11 AM	Science
12 PM	Lunch
1 PM	Sports
2 PM	Reading

Class	Days
Reading	Monday, Wednesday
Math	Tuesday, Thursday
Sports	Monday, Thursday
Science	Tuesday, Friday

1) If Roberts lost his watch during the Sports class, which one of the following must be true?
   A) He lost his watch either in the morning or in the afternoon.
   B) He lost his watch on a Tuesday.

2) If Roberts lost his watch during the Math class, which of the following cannot be true?
   A) He lost his watch in the afternoon.
   B) He lost his watch on a Tuesday or a Thursday.

3) If Roberts lost his watch during the Reading class, which of the following must be true?
   A) He lost his watch either on Monday or on Wednesday.
   B) He lost his watch after noon.

Analytical Reasoning

Name —————————————— Date ——————————

1 GROUPING AND SUMMARIZING

The list below shows the contents of a bag.

Contents	Quantity
Grape	3
Carrot	2
Orange	2
Onion	4
Apple	4
Cauliflower	3

Fill in the grid below that shows vegetables and fruits grouped together.

	Vegetables	Quantity	Fruits	Quantity
Total	Total vegetables		Total fruits	

There are more number of vegetables than there are fruits.
    A) True   B) False

Analytical Reasoning

1	VENN DIAGRAM

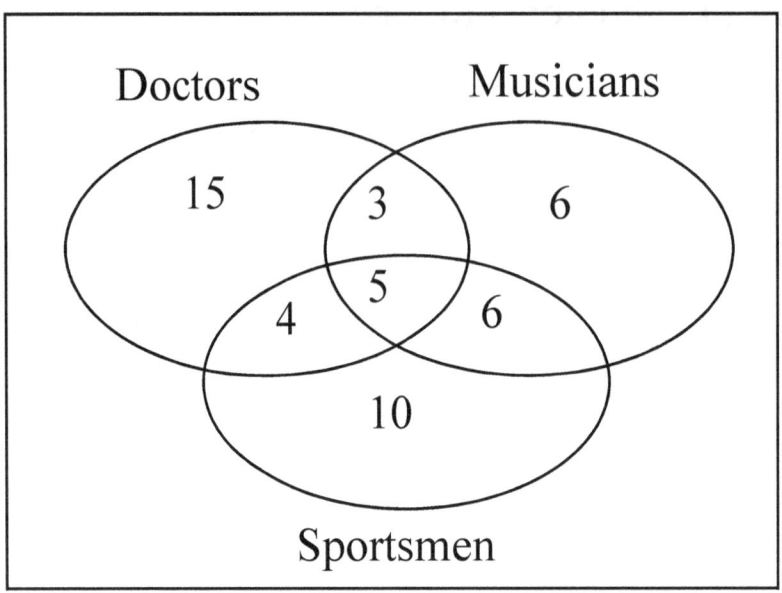

Everyone in a conference room are either doctors, or musicians, or sportsmen, or a combination of these. The Venn diagram shows the number of people belonging to each groups. Answer the following questions based on the information provided.

1) How many doctors are not musicians or sportsmen?

2) How many are sportsmen as well as doctors?

3) How many sportsmen are neither doctors nor musicians?

4) How many people are there in the room?

5) How many are either doctors or musicians?

6) How many are sportsmen and doctors only?

VENN DIAGRAM

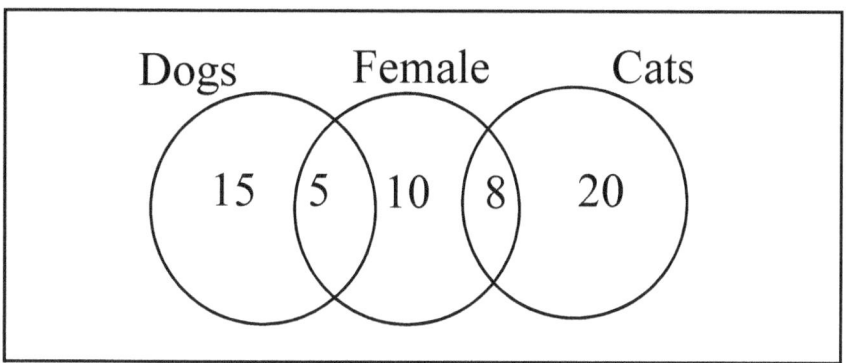

All the animals in a pet care facility can be represented by the Venn diagram shown above. Answer the questions based on the Venn diagram.

1) How many dogs are female?

2) There are no animals other than cats and dogs.
   A) True   B) False

3) How many animals are female?

4) How many male dogs are there in the pet care facility?

5) How many cats are there in the pet care facility?

6) How many animals are in the pet care facility?

3 VENN DIAGRAM

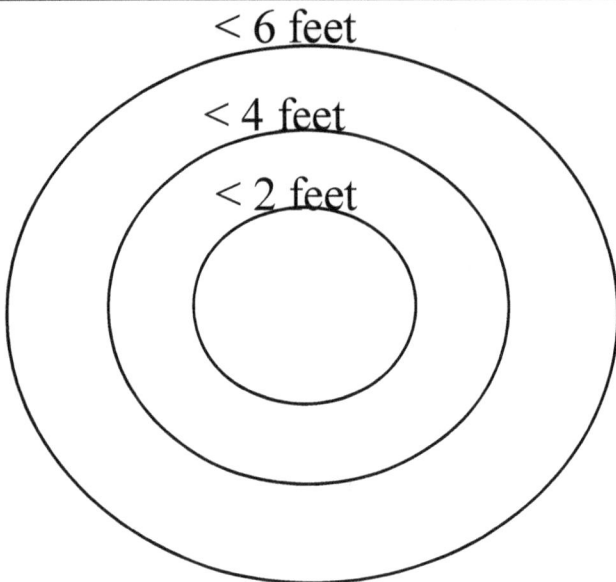

Information about the heights of six people are listed below.
A and B are less than 2 feet.
C and D are more than 2 feet and less than 4 feet.
E and F are more than 4 feet, but less than 6 feet.

Place A,B,C,D,E, and F in the Venn diagram shown above and answer the questions below.

1) Who are less than four feet tall?

2) Who are more than two feet, but less than six feet tall?

3) Who are not less than four feet tall?

4) Who are less than two feet tall or more than 4 feet, but less than 6 feet tall?

Name ——————————————— Date———————————

1 GRAPH LOGIC

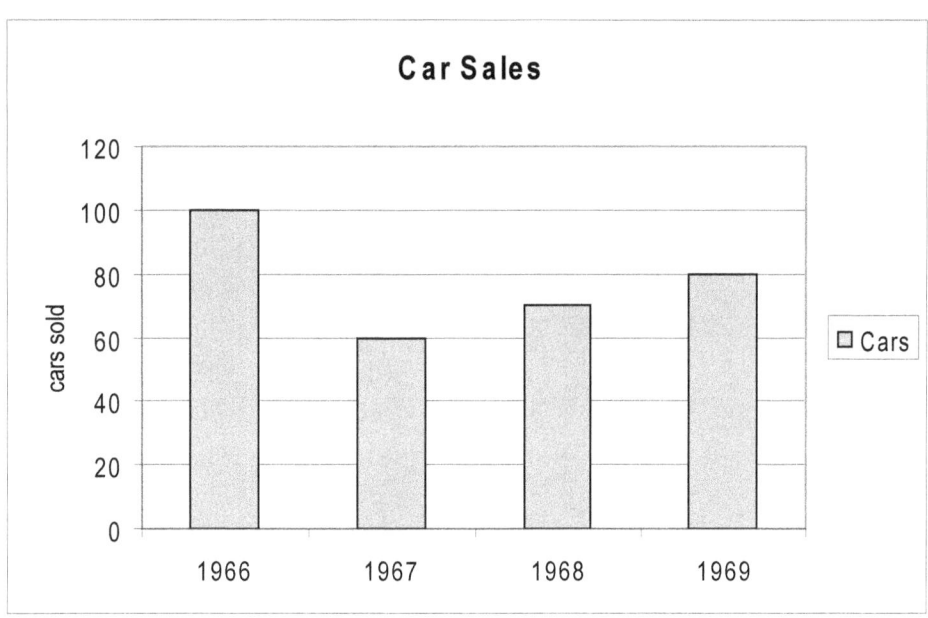

The sales data for a car manufacturing company is shown in the bar graph above. Read the graph and answer the questions below.

1) The one year decline in sales from 1966 to 1967 was less than the one year increase in sales from 1967 to 1968.
    A) True   B) False

2) From 1966 to 1969, there were more growth periods than periods of decline in car sales.
    A) True   B) False

Name —————————————— Date ——————————

2   GRAPH LOGIC

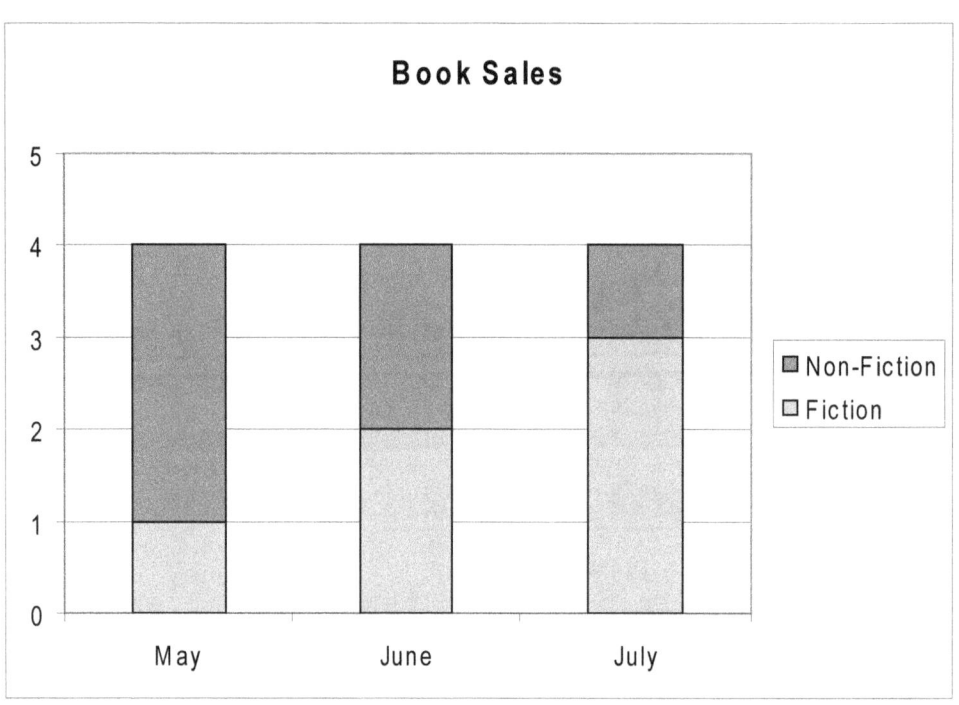

A small book store recorded sales of fictional and non-fictional books as shown above in the stacked bar graph for the months of May, June, and July. Read the graph and answer the questions below.

1) There was a 50% decrease in non-fiction book sales from May to June and June to July.
    A) True   B) False

2) There was a 50% increase in fiction book sales from June to July.
    A) True   B) False

GRAPH LOGIC

A nursery sold flowers in bunches of 50 and 100 and displayed the following symbols.

♀ = 100 flowers

♂ = 50 flowers

John bought several flowers from the nursery and planted them in two gardens, garden-1 and garden-2.

Garden-1 has ♀♀♂ flowers.

Garden 2 has ♂♂♀ flowers.

Answer the questions below based on the information presented above.

1) Garden-2 has more flowers than garden-1.
   A) True   B) False

2) Garden-1 has more 100 flower bunches than garden-2.
   A) True   B) False

4 GRAPH LOGIC

Rainfall Measurements

	1" rain	2" rain	3" rain
October	4	3	1
November	3	4	0
December	0	5	3

The rainfall recording for an area is shown in the chart above. For example, 1" rain was recorded four times in October. Read the chart and answer the questions below. The symbol " refers to an inch.

1) The number of times it rained 2" or below in October is the same as the number of times it rained 3" or below in November.
    A) True   B) False

2) It rained 3" more number of times than it rained 1".
    A) True   B) False

GRAPH LOGIC

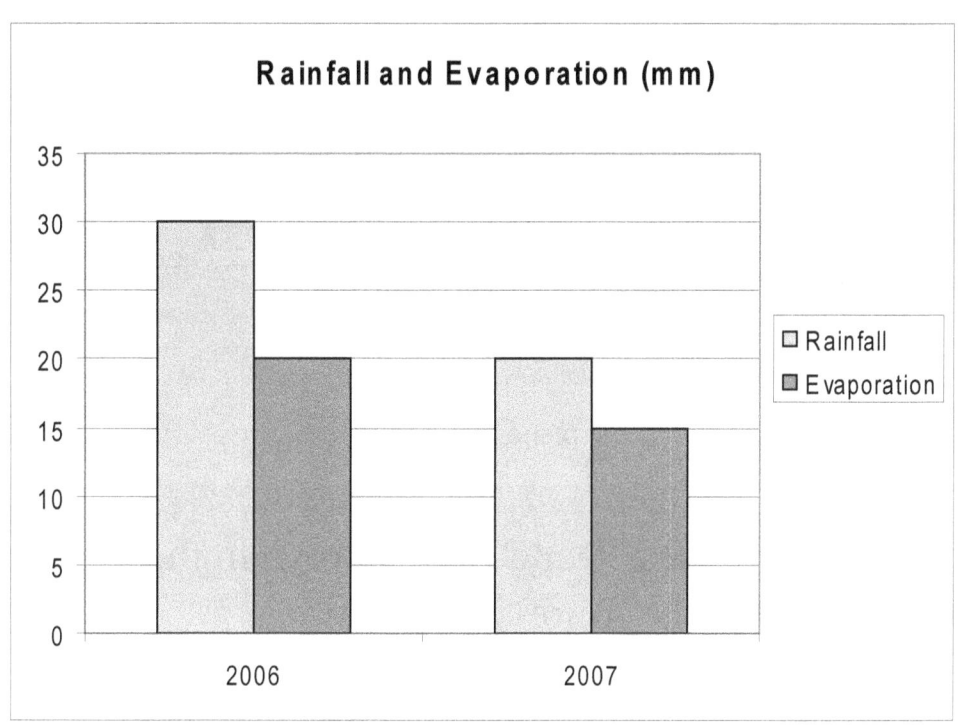

The bar graph show the rainfall and evaporation amounts in millimeters in a certain area for two years. Read the graph and answer the questions.

1) More water evaporated in 2006 than in 2007.
   A) True   B) False

2) More water was retained in 2007 than in 2006.
   A) True   B) False

GRAPH LOGIC

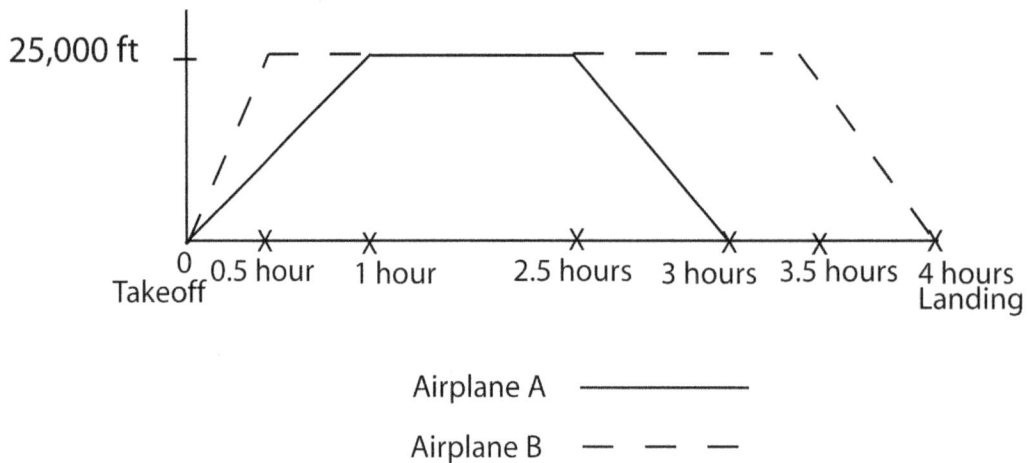

The trajectory of two airplanes, A and B are shown in the graph above. Read the graph and answer the following questions.

1) Airplane A climbed to 25,000 feet faster than airplane B.
   A) True   B) False

2) Airplane B took more time to descend 25,000 feet than airplane A.
   A) True   B) False

3) Both airplanes cruised for the same duration.
   A) True   B) False

GRAPH LOGIC

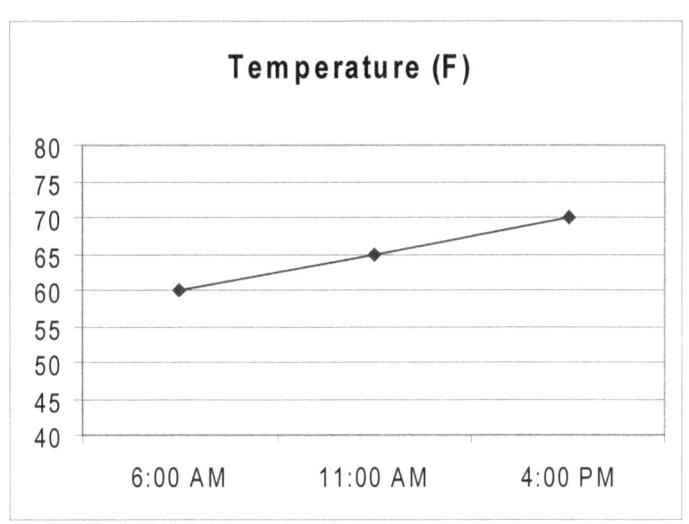

Temperature in city-A rose at a constant rate from 60 degrees at 6 AM to 70 degrees at 4 PM as shown in the line graph above. Read the graph and answer the questions.

1) If the temperature in another city, city-B is 55 degrees at 6 AM and increased at the same rate as city-A, then city-B will be warmer than city-A at 4 PM.
       A) True   B) False

2) If the temperature in city-B is 55 degrees at 6 AM and increased at twice the rate as city-A, then both cities will have the same temperature at 11 AM.
       A) True   B) False

NUMBER LOGIC

Figure out the logic in the sequence and find the missing number.

| 1 | 1.25 | 1.50 | ? | 2 |

| 2 | 6 | 2 | 4/3 | ? |

| 3 | 9 | ? | 8 | 7.5 |

| 4 | 1 | 1/2 | 1/4 | ? |

| 5 | 8 | -8 | 16 | ? |

| 6 | 10 | -10 | 20 | ? |

| 7 | 1 | 2 | 1 | 3 | 1 | 4 | ? |

Name _____   Date _____

NUMBER LOGIC

Figure out the logic in the numbers and find the missing number.

8
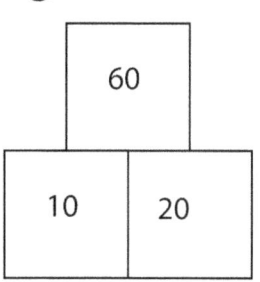 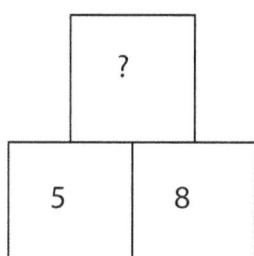

9
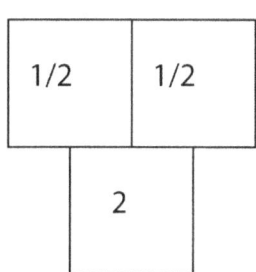 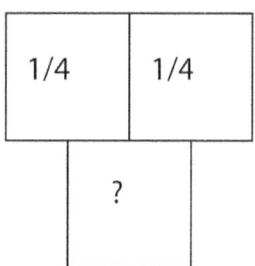

10
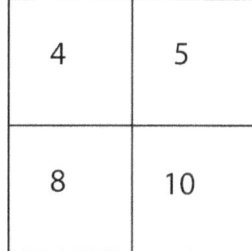 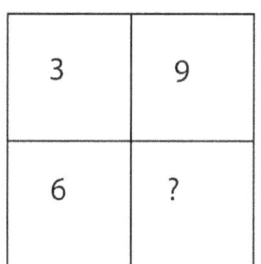

11
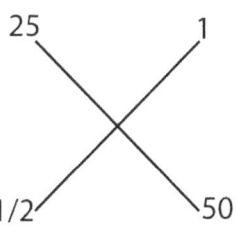 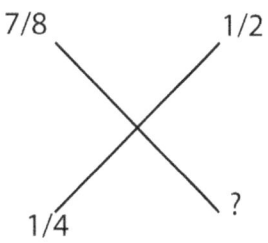

12
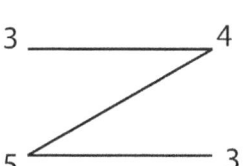

Name —————————————— Date ——————————

NUMBER LOGIC

Figure out the logic in the numbers and find the missing number.

13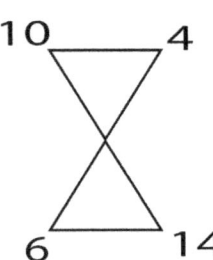

14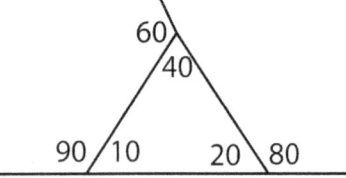

15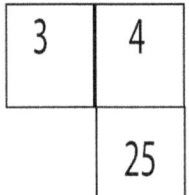

16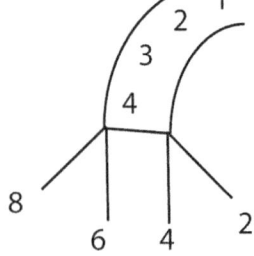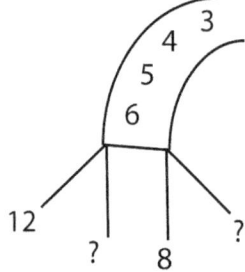

LETTER LOGIC

Figure out the logic in the sequence and find the missing letter or number.

1. A21   B63   ?

2. 3C3   4D4   ?

3. 26-Z   1-A   25-Y   ?

4. P^Q   Q^R   ?

5. T/V   V/X   X/?

6. BCD   24   EFG   ?

7. BDF   12   CEG   15   MOQ   ?

Name —————————————— Date ——————————

LETTER LOGIC

Figure out the letter or number indicated by the question mark?

8

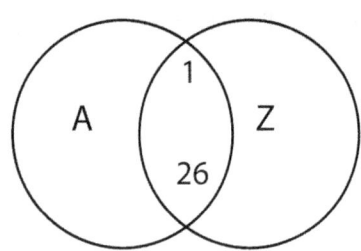 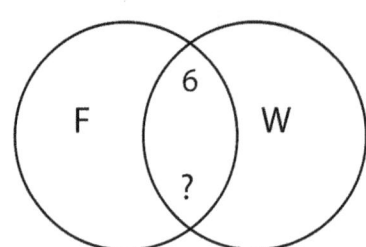

9

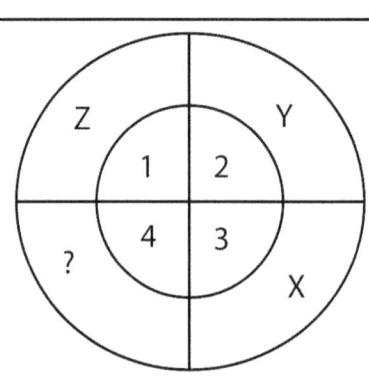

10

L	M	N
N	O	?

11

A	E	I
C	G	?
E	I	M

12

If A=1, B=2 and so on, then JACK is
    A) 110110   B) 101311   C) 110311

If A=1, B=2 and so on, then LAD is
    A) 1412 B) 4112 C) 1214

LETTER LOGIC

13
Figure out the logic in the sequence and find the missing letter or number.

1, 2, C   5, 6, G   8, 9, ?

14
If G is A and D=M then GOOD =
   A) AOOM   B) GOOM   C) AOOD

15
Aliens visiting earth talk in a language similar to English. When they say "RAC", they actually mean "CAR" and when they say "ESOUM", they actually mean "MOUSE".

1) What do the aliens mean when they say TRAMS?
   A) TRAIN   B) TRAM   C) SMART

2) What do the aliens mean when they say CIGOL?
   A) CIGAR   B) LOGIC   C) ROGER

3) What do the aliens say when they want to refer to a BOY?
   A) OYB   B) YOB   C) BYO

1 SUDOKU

Solve the following Sudoku. A correctly solved Sudoku has numbers 1-9 appearing only once in each row, each column and each 3x3 grid. You gain valuable positioning skills by solving these sudokus.

5		8	3		1	4		9
9	3		6	5		1	2	
6		2	4	9	8	5	7	
8		5	7		3	2		6
3	4			1		8		5
	2	6		8	5		3	4
2		1	8		9	6		7
	8	9	1		4		5	2
4	6		5	7	2		8	1

Analytical Reasoning   Answers-119   62
© Gift Of Logic, Inc * Copying prohibited

2 SUDOKU

Solve the following Sudoku. A correctly solved Sudoku has numbers 1-9 appearing only once in each row, each column and each 3x3 grid. You gain valuable positioning skills by solving these sudokus.

1	9	7	8	5	6	4	3	2
	6	8	2		4	7	9	1
2	4	3		9	1		5	6
9	3	2	6	1	8	5	4	7
7	5	4	3	2	9	6	1	8
8		6	4		5	9		3
6	8	9	1	4	3	2	7	5
4	7	1	5	6	2	3	8	9
3	2		9	8		1	6	

Name _____ Date _____

3
SUDOKU

Solve the following Sudoku. A correctly solved Sudoku has numbers 1-9 appearing only once in each row, each column and each 3x3 grid. You gain valuable positioning skills by solving these sudokus.

3	1	2	8		4	6		7
7		9	6	2	1	8	4	3
6	8	4	9	7		1		5
	7	6	4	3	8	9	1	2
9	3		5		2	4	7	8
2	4	8	7	1	9		3	
8		3	1		7	2	5	
4	2		3	9	6	7	8	1
1	9	7	2	8	5	3	6	4

Name ______________________ Date______________________

4

SUDOKU

Solve the following Sudoku. A correctly solved Sudoku has numbers 1-9 appearing only once in each row, each column and each 3x3 grid. You gain valuable positioning skills by solving these sudokus.

9	8	4		6	5	3		2
6	5		9		1	8	4	
	1	3	8	4	2	9	5	6
8	3	6		5		2	7	9
4		5	2		3	1	6	
2	7		6	9	8	4	3	5
1	6	8	3	2	7	5	9	4
3		9	5		6	7	2	
5	2	7		1	9		8	3

Name _____ Date _____

5
SUDOKU

Solve the following Sudoku. A correctly solved Sudoku has numbers 1-9 appearing only once in each row, each column and each 3x3 grid. You gain valuable positioning skills by solving these sudokus.

5	7	8	6		3	1	2	9
1		6	8	7	9	4		3
4	9		2		1	7		8
6	8	9	1	3	5		7	4
		5	4		7	9	8	6
2	4	7	9	8		5		1
9	6		7	1	8	3	4	5
	5	4		9	2	6	1	
7		1	5	6		8	9	2

Analytical Reasoning Answers-123
© Gift Of Logic, Inc * Copying prohibited

6
SUDOKU

Solve the following Sudoku. A correctly solved Sudoku has numbers 1-9 appearing only once in each row, each column and each 3x3 grid. You gain valuable positioning skills by solving these sudokus.

6		1	7		2	4		9
5	4	9	3	8	1	7	6	
2	8	7	4		6	1		5
9		4	8	6	7	3	2	1
	2	8		1	9			4
7	1	6	2	3	4	9	5	
4	7		1	2		5	9	
1		2	9		5	8		3
8	9	5		4	3	2	1	7

Name ——————————————— Date ————————————

PICTORIAL REASONING

Name _____   Date_____

PICTURE SEQUENCE

Figure out the logic in the picture sequence, and draw the next picture in the sequence.

1

 ?

2

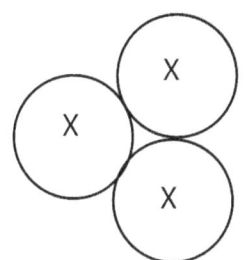   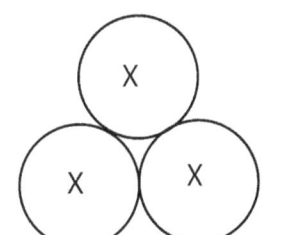   ?

3

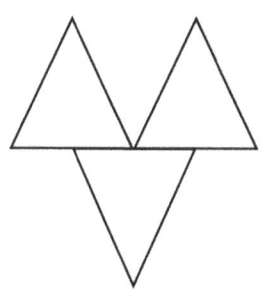   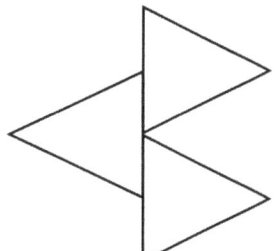   ?

4

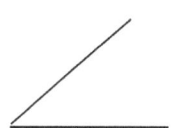      ?

Pictorial Reasoning   Answers-125
© Gift Of Logic, Inc * Copying prohibited

Name —————————————— Date ————————————

PICTURE SEQUENCE

Figure out the logic in the picture sequence, and draw the next picture in the sequence.

5  ?

6  ?

7 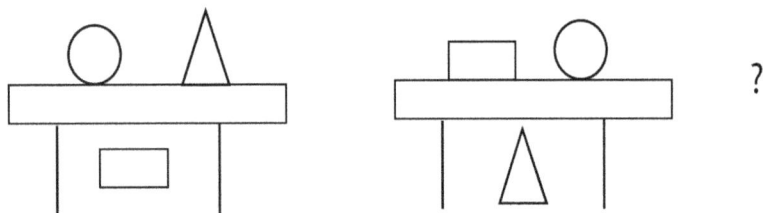 ?

8 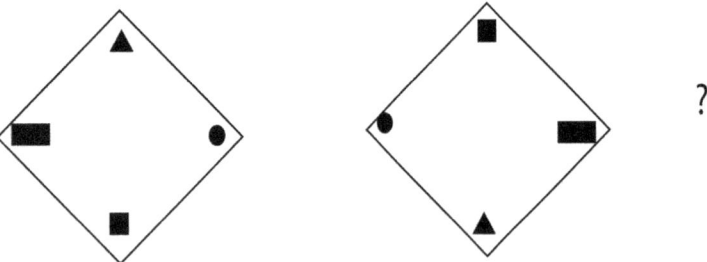 ?

9 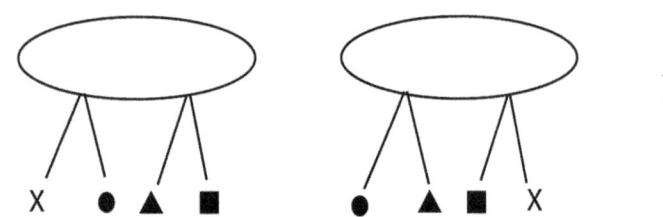 ?

Pictorial Reasoning   Answers-126   70
© Gift Of Logic, Inc * Copying prohibited

Name ——————————   Date——————

PICTURE SEQUENCE

Figure out the logic in the picture sequence, and draw the next picture in the sequence.

10 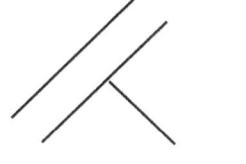   ?

11 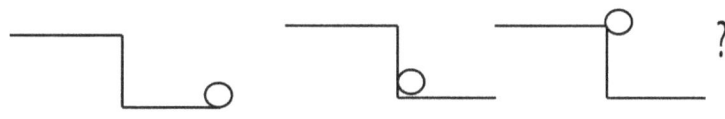   ?

12 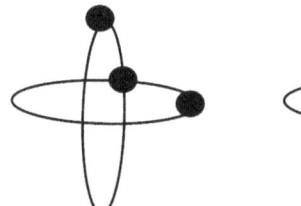   ?

13 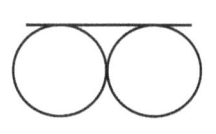   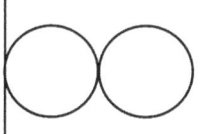   ?

14 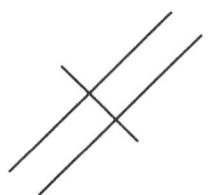   ?

Pictorial Reasoning   Answers-127
© Gift Of Logic, Inc * Copying prohibited

Name _____   Date _____

PICTURE ANALOGY

Figure out the logic in the picture analogy, and draw the correct picture that will complete the analogy.

1

2

3

4

5

Pictorial Reasoning   Answers-128   72
© Gift Of Logic, Inc * Copying prohibited

PICTURE ANALOGY

Figure out the logic in the picture analogy, and circle the correct picture that will complete the analogy.

6 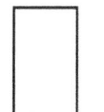 : 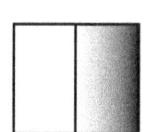 AS  : A  B

7  : 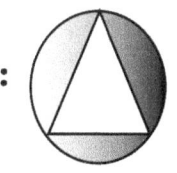 AS 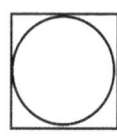 : A 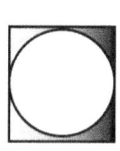 B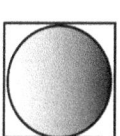

8  : AS  : A  B

9 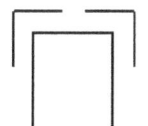 : 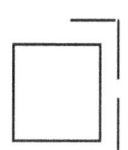 AS 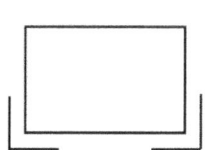 : A 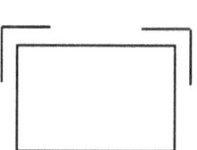 B

Pictorial Reasoning

Name _____   Date _____

PICTURE ANALOGY

Figure out the logic in the picture analogy, and circle the correct picture that will complete the analogy.

10   A   B   C

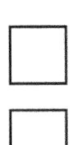 : 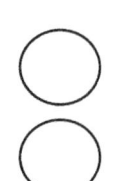 AS  :

---

11   A   B   C

 AS 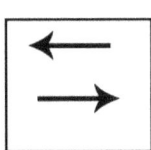 :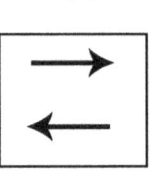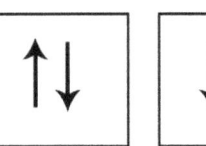

---

12   A   B   C

 :  AS 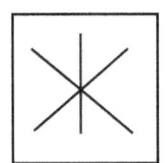 :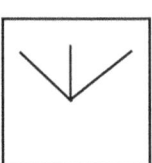

---

13   A   B   C

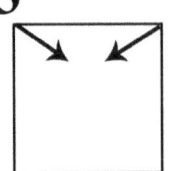 : 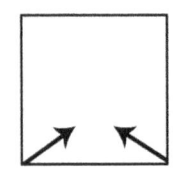 AS 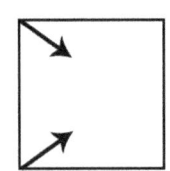 :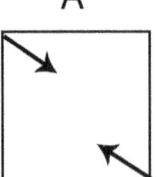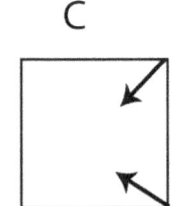

---

14   A   B

 AS 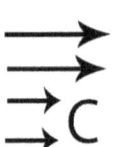 :

Pictorial Reasoning   Answers-130
© Gift Of Logic, Inc * Copying prohibited

Name _____   Date _____

ODD PICTURE

Figure out the logic in the pictures, and identify the odd picture.

1 A   B   C
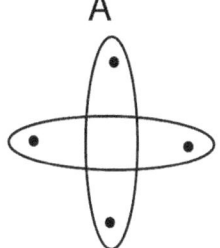 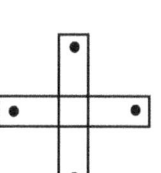 

2 A   B   C
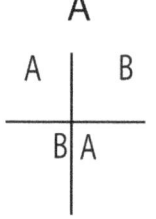 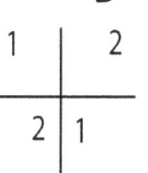 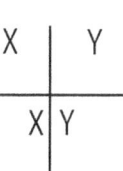

3 A   B   C
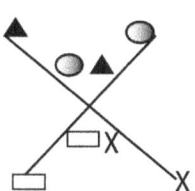 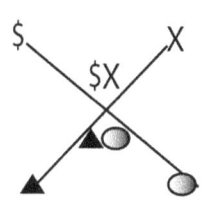 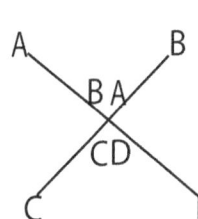

4 A   B   C
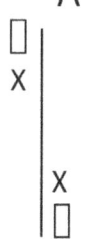  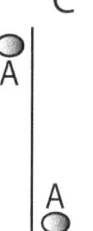

5 A   B   C
 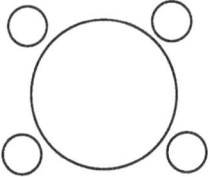 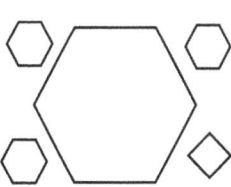

Pictorial Reasoning   Answers-131

© Gift Of Logic, Inc * Copying prohibited

Name —————————————— Date ——————————

ODD PICTURE

Figure out the logic in the pictures, and identify the odd picture.

6   A   B   C

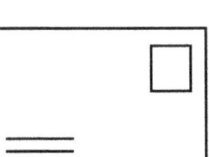 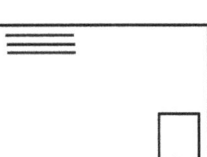 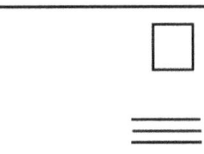

7   A   B   C

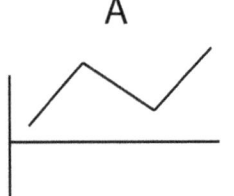 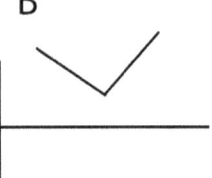 

8   A   B   C

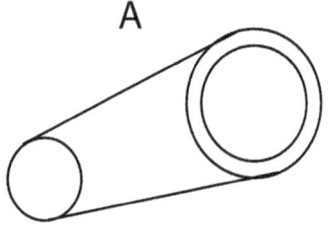 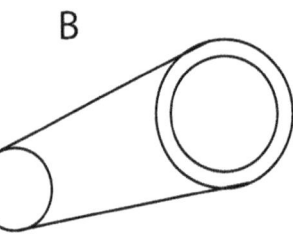 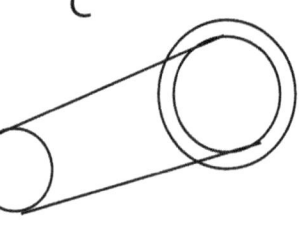

9   A   B   C

10   A   B   C

 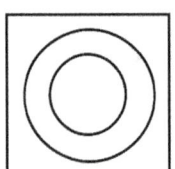 

Pictorial Reasoning   Answers-132

© Gift Of Logic, Inc * Copying prohibited

Name —————————————— Date————————

PICTURE DIFFERENCE

Mark the differences in the set of pictures shown, with arrows.

1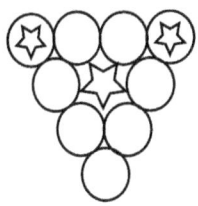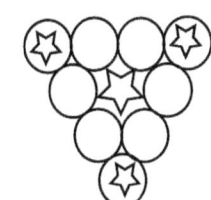

2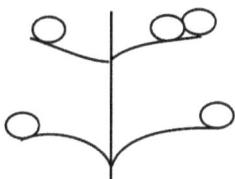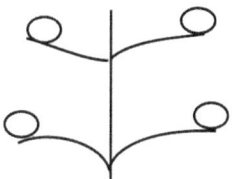

3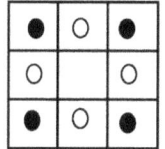

4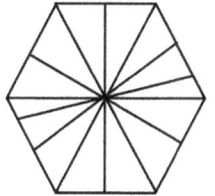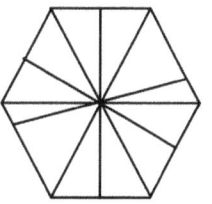

5

1	2	3
4	5	6
7	8	9
0	*	#

1	2	3
4	5	6
7	8	9
#	*	#

Pictorial Reasoning

Name _____   Date _____

PICTURE DIFFERENCE

Mark the differences in the set of pictures shown, with arrows.

6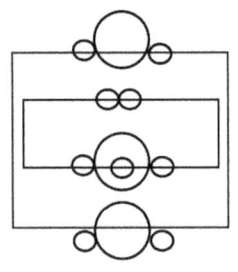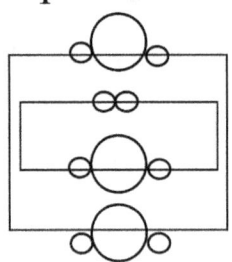

7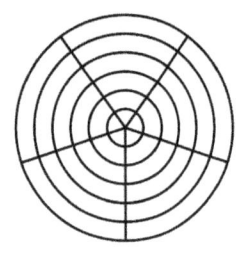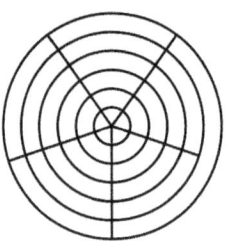

8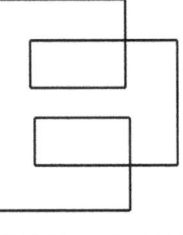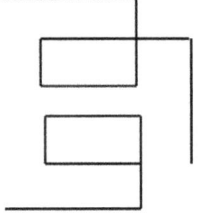

9

10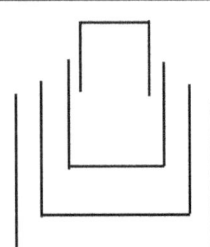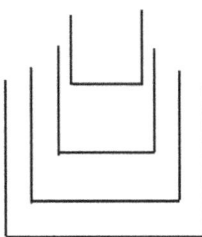

Pictorial Reasoning   Answers-134

© Gift Of Logic, Inc * Copying prohibited

Name _____ Date_____

PICTURE DIFFERENCE

Mark the differences in the set of pictures shown, with arrows.

11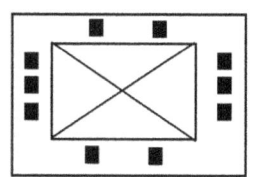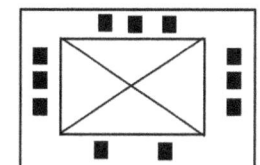

12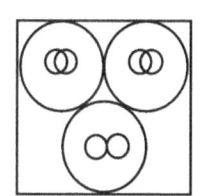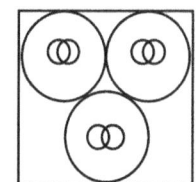

13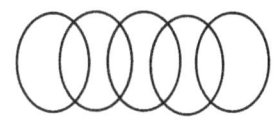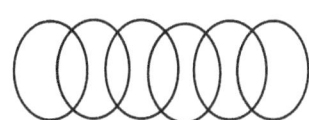

14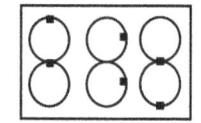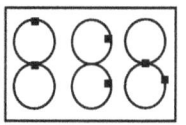

15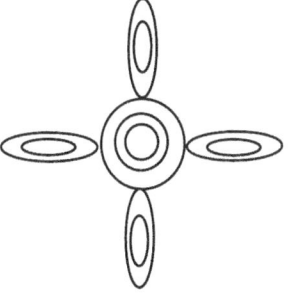

Pictorial Reasoning   Answers-135
© Gift Of Logic, Inc * Copying prohibited

Name —————————————— Date ———————————

PATTERN MATCHING

Find the logical pattern in the pictures on the left, and identify the picture on the right that will fit in the space marked with ? to complete the pattern.

1
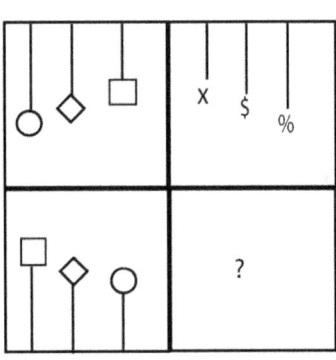   A 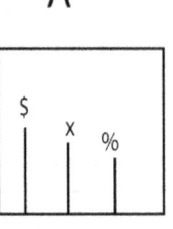   B 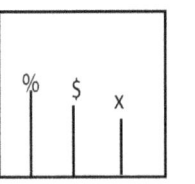   C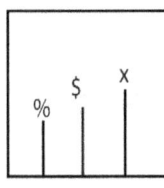

2
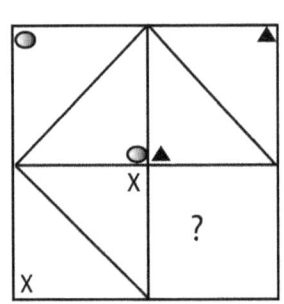   A 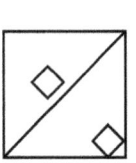   B 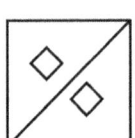   C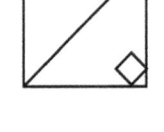

3
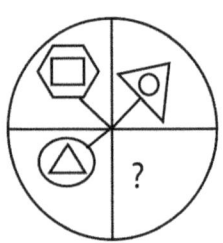   A   B   C
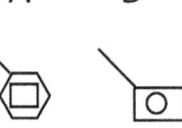 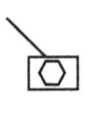

4
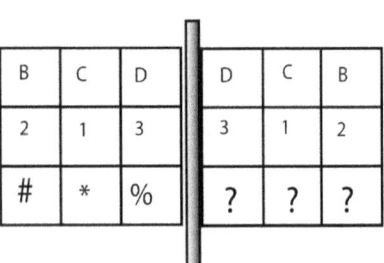   A   B   C
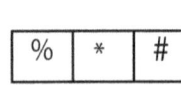

Pictorial Reasoning   Answers-136
© Gift Of Logic, Inc * Copying prohibited

ANSWERS

INFERENCING - Only If, If and only if

1 Trash will be picked up only if it is a Friday.

Answer: B) If it is not a Friday, then trash will not be picked up.

Reasoning: The conditional statement can be written in symbolic form as follows: Trash → Friday   Note the antecedent and the consequent. The contrapositive is   ~Friday → ~Trash. So, choice B correct.

Choice A is the converse and is therefore not true. Choice C is the inverse of the conditional and is also not true.

---

2 Sonya will sing if and only if Tonya sings.

If the above information is true, which one of the following <u>must be true</u>?

Answer: C) If Tonya sings, then Sonya will sing.

Reasoning: Note the biconditional "if and only if". This statement can be represented as Sonya ↔ Tonya. This can be broken down further as follows: Sonya → Tonya and Tonya → Sonya. The contrapositives are true and they are: ~Tonya → ~Sonya and ~Sonya → ~Tonya

So, choice C is the correct answer. Choices A and B are incorrect since they do not match the contrapositives.

Answers

INFERENCING - Categorical

Venn diagrams for questions 1-3 are shown at the bottom.

1 All Italians are Europeans.

1) If you are an Italian, you are an European. Answer: A) Valid. The circle for Italians is inside the circle for Europeans.

2) If someone is an European, that person is an Italian. Answer: B) Invalid. We can infer from the Venn diagram that there can be Europeans who are not Italians - this is shown by "x" in the Venn diagram.

2 No criminal is innocent.

1) There are some innocent people who are criminals. Answer: B) Invalid If this were the case, then the two circles would have some members in common.

2) There is no innocent person who is a criminal. Answer: A) Valid This is the converse of the given statement.

3 Some clowns are students.

1) Some students are clowns. A) Valid. The intersecting portion of the Venn diagram shows that some students are clowns.

2) All students are clowns. Answer: B) Invalid. In the diagram, we can have "y", a student who is not a clown. So, we cannot infer this.

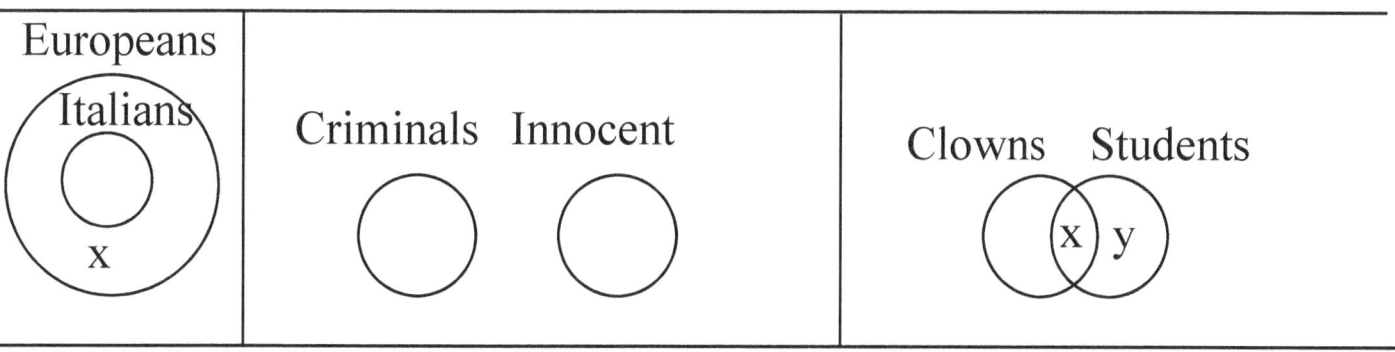

Answers
© Gift Of Logic, Inc * Copying prohibited

INFERENCING - Categorical

4   Some rats have long tails.

1) Some animals that have long tails are rats. Answer: A) Valid. "x" in the Venn diagram shows an animal with a long tail that is also a rat.

2) All animals that have long tails are rats. Answer: B) Invalid. As shown in the Venn diagram, "y" is an animal with a long tail, but it is not a rat.

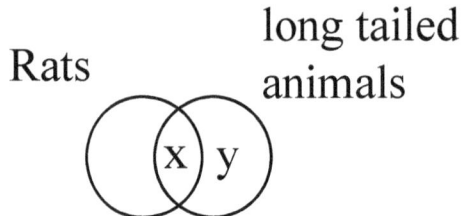

5   All politicians are human beings.

1) If there is a human being, he must be a politician.
    Answer: B) Invalid. The "x" in the Venn diagram shows a human being who is not a politician.

2) There is no human being who is also a politician.
    Answer: B) Invalid. Since all politicians are human beings, this statement is invalid. The "y" in the Venn diagram represents a politician who is also a human being.

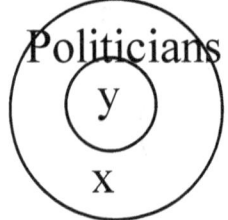

INFERENCING - Categorical

6

All the locks can be opened with this key. Bobby has a lock.

If the above statements are true, which of the following cannot be true?
Answer: B) Bobby's lock cannot be opened with this key.

Reasoning: This is clear since "all" the locks can be opened with this key. So, Bobby's lock can also be opened with this key.

7

All the boys must stand behind the girls. Chad is a boy.

If the above statements are true, which of the following cannot be true?
Answer: B) A girl stood behind Chad.

Reasoning: Since Chad is a boy and "all" the boys must stand behind the girls, a girl cannot stand behind Chad.

8

No girl can be in the sports team. Angelina is a girl. Brad is a boy.

If the above statements are true, which of the following cannot be true?
Answer: A) Angelina is in the sports team.

Reasoning: Angelina cannot be in the sports team because she is a girl and "no girl" can be in the sports team.

Answers

INFERENCING - Categorical

9

Some reptiles are dangerous. Some are not. The ones that are not dangerous make good pets. The lizard is not a dangerous reptile.

If the above statements are true, which of the following cannot be true?
    Answer: B) Lizards do not make good pets.

Reasoning:
    not dangerous → makes good pet
    lizard is not dangerous
    so, the lizard will make a good pet.
To say that Lizards do not make good pets cannot be a true statement.

10

No student can use a cell phone in the class. Anita is a student.

If the above statements are true, then which of the following cannot be true?
Answer: A) Anita can use her cell phone in the class.

Reasoning:
    student → ~use cell phone
    Anita is a student
    so, Anita cannot use the cell phone

To say that Anita can use her cell phone cannot be a true statement.

Answers
© Gift Of Logic, Inc * Copying prohibited

CAUSE AND EFFECT

In the following table, a few causal statements are shown with their cause and effect. Complete the rest.

Causal Statement	Cause	Effect
P c→Q	P	Q
Rain causes flooding.	Rain	Flooding
Flooding is caused by rain.	Rain	Flooding
Flooding is the effect of rain.	Rain	Flooding
The reason that we lost the game is due to poor team spirit.	Poor team spirit	lose the game
Heavy load on his shoulders is responsible for his backache.	heavy load	backache
The effect of global warming is unpredictable weather.	global warming	unpredictable weather
The rabbit chewed the plant causing it to die.	rabbit chewing	plant dies
Janet is having health problems because of overeating.	overeating	health problems
Tom's carelessness caused the boat accident.	Tom's carelessness	boat accident

Answers

INFERENCING - Causal

In the following table, a few causal statements along with some inferences are shown. Evaluate the inferences as valid or invalid.

Causal statement	Inference	Valid/Invalid
The ball landed on the table causing the milk to spill. The milk that spilled spoiled the carpet.	The ball that landed on the table caused the carpet to be spoiled.	Valid. The root cause "caused" the effect.
A fire in the building caused the firefighters to come.	The firefighters caused the fire in the building.	Invalid. The converse of a causal cannot be inferred.
First there was an explosion, then there was smoke.	The explosion caused the smoke. The smoke caused the explosion.	Both are invalid. Succession of events do not imply causation.
Tom pushed Harry and Harry fell on a lamp. So the lamp broke.	Tom's push caused the lamp to break.	Valid. Tom's push was the root cause.
Pizza is tasty. Pizza is fatty.	The fat in the pizza causes the taste.	Invalid. Correlation does not imply causation.
High temperatures will cause power outage. The temperature today is very high.	There will be power outage today.	Valid high temp c→ power outage. Temp today is high. So, there will be power outage.

Answers

INFERENCING - Causal

1 Riding bikes regularly causes..

Answer: A) Rosie has strong leg muscles.

Reasoning: Note the causal relationship mentioned here.
   regular bike riding c→ strong leg muscles.

Using this causal relationship, we can infer that since Rosie rides her bike regularly, she will have strong leg muscles.

---

2 While it is true that there was a mild earthquake..

Answer: B) If the quality of the foundation was not poor, the building would not have collapsed.

Reasoning: Note the causal relationship here.
   poor quality of foundation c→ building collapse

So, since the poor quality of the foundation was the cause, fixing this cause will prevent the effect.

Choice A is not correct because even if the earthquake had not happened, the poor quality of the foundation could have possibly led to the collapse of the building.

Answers
© Gift Of Logic, Inc * Copying prohibited

AGREE/DISAGREE

In questions 1-8, read the information given and agree or disagree with it as instructed.

1 Recycling is good for the environment. So, newspapers must not be thrown into the trash bin, but instead, they must be recycled.

Agree: Recycling newspapers is a good thing since it will help save a lot of trees from being cut.

2 Don't paint the wall blue. There are blue colored pictures hanging on the wall and they won't look nice on a blue wall.

Disagree: The blue colored pictures have a red frame. So, the color of the wall does not matter.

3 Walking to school is good for our health. Moreover, by not taking our cars and motorbikes to school, we can help reduce air pollution. Therefore, we all should walk to school as much as possible.

Agree: Cars and motorbikes pollute a lot. By walking to school, we can cut down air pollution. Plus, walking is good for our heart. So, I agree that we must try to walk to school as much as possible.

Answers

© Gift Of Logic, Inc * Copying prohibited

AGREE/DISAGREE

4 We must turn off the lights when we do not need them.

Agree: I agree. When the lights are not needed, it is better to turn them off and save some energy.

---

5 We should maintain a good landscape in front of our house. This will make our house look beautiful.

Agree: Planting seasonal flowers in the flower bed and having a well maintained lawn definitely will make our house look nice.

---

6 The ball that he threw might have hit the bulb. That is why the bulb fell down and broke into pieces.

Disagree: I disagree. The bulb was loose and broke even before the ball hit the bulb.

---

7 I think Madrid is the capital of Egypt.
Disagree: Madrid is the capital of Spain, not Egypt. Cairo is the capital of Egypt.

---

8 Doctors work very long hours. This is not good because they can make mistakes. So, doctors must not work very long hours.

Disagree: Doctors are trained to work long hours without making mistakes. Moreover, they work with other medical personnel who will assist them. So, doctors can work very long hours without any problem.

SENTENCE ANALOGY

1 Statement: Our stomach is like a balloon.

What are being compared?   stomach and balloon

Answer: A) As we eat more, our stomach gets bigger.
The balloon gets bigger when it is filled with air. The stomach gets bigger when it is filled with food. Choice B is incorrect as it goes against the analogy.

2

Statement: We grow in height the same way as trees grow in height.

What are being compared in the statement?
         our growth and the growth of a tree

Answer: B) We grow tall for some time and then stop growing.
Trees also grow tall for some time and then stop growing.

---

3   Statement: The purpose of our eyelids is similar to the purpose of an umbrella.

What are being compared in the statement? eyelids and umbrella

Answer: A) Eyelids protect the eyeballs.
An umbrella protects whatever is below it from damage. The eyelids protect the eyeballs that are below it.

SENTENCE ANALOGY

4 The relation between Tom and Dick is similar to the relation between two adversaries.

What are being compared in the statement?
Relation between Tom and Dick and the relation between two adversaries.

Answer: B) Tom and Dick often quarrel with each other.
Adversaries are opponents, and they quarrel with each other.

---

5 This problem is a tough nut to crack.

What are being compared in the statement?
A problem and a tough nut

Answers: B) This problem is difficult to solve.
Tough nuts are difficult to crack. Similarly, a tough problem is difficult to solve.

---

6 Holly dresses like a fashion model.

What are being compared in the statement?
Holly and a fashion model.

Answer: A) Holly's dresses are contemporary.
Fashion models wear dresses that are of modern style (contemporary). They are not outdated.

WORD ANALOGY

Q#	Ans	Reasoning
1	B	Spain is to Madrid as England is to London
2	C	Sound is to silence as Light is to darkness
3	B	Bird is to parrot as Animal is to fox
4	C	Man is to boy as Woman is to girl
5	C	Pilot is to turbulence as Sailor is to gust
6	C	Plumber is to pipe as Carpenter is to wood
7	C	Lawyer is to court as Doctor is to hospital
8	B	January is to December as Sunday is to Saturday
9	C	First is to last as Begin is to end
10	C	introduce : recall :: insert : retract
11	C	brain : think :: stomach : digest
12	B	square : rectangle :: circle : ellipse
13	B	eye : vision :: ear : hearing
14	B	lawyer : paralegal :: doctor : nurse
15	B	virus : sickness :: earthquake : destruction
16	C	whole : part :: earth : continent
17	C	part : whole :: classrooms : school
18	C	book : page :: page : sentence

Answers

1 LIST PROCESSING - alphabetic sorting

Sort the following list of names in ascending and descending order.

Name	Ascending	Descending
Steve	Arjun	Zita
Preety	Bobby	Steve
Bobby	Huy	Preety
Huy	Karan	Karan
Karen	Preety	Huy
Arjun	Steve	Bobby
Zita	Zita	Arjun

1) Who is the first in the list when sorted in the ascending order?
    Arjun

2) Who is the first in the list when sorted in the descending order?
    Zita

3) When sorted alphabetically in ascending order, Zita is the last on the list. So, Zita is the tallest in the group.
    Answer: B) Invalid. The sort is based on the first letter in their names, not on their height.

5) When a list that has an odd number of items is sorted in ascending order and descending order, one item in the list will have the same rank.
    Answer: A) True. For example, the above list has 7 names and Karan has the fourth rank in both the lists. If there were 9 names, then the 5th name will have the same rank in the ascending and descending order sorts.

Answers
© Gift Of Logic, Inc * Copying prohibited

2   LIST PROCESSING - numeric sorting

Sort the ID# column in ascending and descending order and write it in the respective columns.

ID#	Ascending Order	Descending Order
10	1	10
7	4	9
8	6	8
6	7	7
1	8	6
4	9	4
9	10	1

Exits in a highway are numbered in descending order going north and in ascending order going south.

1) Harrison was driving his car north towards the mountain and the highway exits he saw were numbered 1, 2 and 3.

This information Answer: B) cannot be true. Numbers 1, 2 and 3 are not in descending order. They must be in descending order going north.

2) Julia was going south in the highway to her grandmother's home and noticed exits numbered 11, 12 and 13.

This information Answer: A) must be true. Going south, the numbers are in ascending order.

Answers
© Gift Of Logic, Inc * Copying prohibited

3 LIST PROCESSING - mixed sorting

Name	Age
Steve	7
Preety	6
Bobby	5
Huy	6
Karan	7
Arjun	5
Zita	6

Sort by Name

Name	Age
Arjun	5
Bobby	5
Huy	6
Karan	7
Preety	5
Steve	7
Zita	6

Sort by Age

Name	Age
Bobby	5
Arjun	5
Preety	6
Huy	6
Zita	6
Steve	7
Karan	7

Answers
© Gift Of Logic, Inc * Copying prohibited

LIST PROCESSING - mixed sorting (continued)

Sort the list shown above in ascending order by age and if there is a tie, sort by name. Write the sorted names below.

Sort by Age and Name

Name	Age
Arjun	5
Bobby	5
Huy	6
Preety	6
Zita	6
Karan	7
Steve	7

Use the lists shown above and answer the following questions.

1) How many people are 6 years old? 3
Which of the lists were the most useful to answer this question and why?
The Sort by Age and Sort by Age and Name lists have the names sorted based on age. So, it is easy to find out how many people are 6 years old from these lists.

2) Why is the Sort by Age list different from the Sort by Age and Name list?
The Sort by Age list sorts only by age. So, if two people have the same age, it does not matter who is first in the list. The Sort By Age and Name list sorts by Age first. If there is a tie, it sorts by Name. For example, for age 5, there are two people - Bobby and Arjun. But, since A comes before B, Arjun is listed before Bobby.

Answers
© Gift Of Logic, Inc * Copying prohibited

4 LIST PROCESSING - SORTING AND RANKING

After sorting a list, it is sometimes necessary to rank the items in the list.

Student	Score
Cynthia	45
Lauren	35
Aparna	20
Omar	40
Josh	30
Arti	25

Sort the list in descending order based on the student scores and write their ranks.

Student	Score	Rank
Cynthia	45	1
Omar	40	2
Lauren	35	3
Josh	30	4
Arti	25	5
Aparna	20	6

Answers

SEQUENCING

1 The following paragraph describes a series of events all jumbled up. Rewrite the paragraph by arranging them in a logical sequence

Following the Math class was the Science class..

The first class in the morning was a Math class. Following the Math class was the Science class. The last class of the day was a Sports class.

---

2 Bill has an appointment with the dentist in the morning. After seeing the dentist, he will go and get groceries. After that, but before going home, he will buy a newspaper.

Which of the following indicates the correct chain of stops planned by Bill
Answer: B) Dentist-Groceries-Newspaper-Home

---

3 Jill went to the park yesterday. Day before yesterday she went to a movie. Tomorrow, she will go to the Zoo. Day after tomorrow she will go to see her grandma. But, today she is at home doing her homework.

Which of the following describes the sequence of places where Jill would be present?

Answer: B) Movie,Park,Home,Zoo,Grandma

Note that the events are described relative to today using day-specific words such as yesterday, day before yesterday, tomorrow, and day after tomorrow.

Answers

© Gift Of Logic, Inc * Copying prohibited

SEQUENCING

4 Sequence based on time

The space shuttle was moved to the launching pad..

Which of the following indicates the times at which events happened until lift off?
Answer: B) 8:00 AM - 9:00 AM - 10:00 AM - 11:00 AM
Note the word "until" in until lift off. So, the timeline must end with the lift off.

---

5 Sequence based on day

Jack has a fitness class every other day..

1) Which one of the following represents the sequence of days when Jack has a fitness class? Answer: B) Monday, Wednesday, Friday. Skip one day starting from Monday.

---

6 Sequence based on recurring events

Pizza is served for lunch every other day ..

Which one of the following represents the sequence of items served during the week starting with Monday.
    Answer: B) Pizza, Sandwich, Pizza, Sandwich, Pizza

Answers

1 CORRELATION

Correlate the information in the tables and answer the following questions.

ID	Name
1	Jeff
2	Chen
3	Prem
4	Gonzales

ID	City	Country
1	Dallas	USA
2	Shanghai	China
3	Chennai	India
4	Madrid	Spain

1) In which city does Prem live? Chennai
2) Who lives in Dallas? Jeff
3) What is the ID Number of the person living in Madrid? 4
4) Who lives in China? Chen

Place all the information in one table shown below.

ID	Name	City	Country
1	Jeff	Dallas	USA
2	Chen	Shanghai	China
3	Prem	Chennai	India
4	Gonzales	Madrid	Spain

Answers

2 CORRELATION

Correlate the information in the following two tables and answer the questions below.

Time	Class
8:00 AM	Craft
10:00 AM	Drawing
2:00 PM	Painting
4:00 PM	Puppetry

Time	Teacher	Room
8:00 AM	Osborne	101
10:00 AM	Gupta	144
2:00 PM	Sweeney	135
4:00 PM	Gilbert	128

1) Name the teachers who will teach in the mornings.
    Osborne, Gupta
2) Name the teachers who will teach in the afternoons.
    Sweeney, Gilbert
3) In which rooms will morning classes be held?
    Rooms 101 and 144
4) In which rooms will the afternoon classes be held?
    Rooms 135 and 128
5) Name the teacher who will teach the Painting class.
    Sweeney
6) What class will Mr. Gilbert teach in room# 128?
    Puppetry

Answers
© Gift Of Logic, Inc * Copying prohibited

3   CORRELATION

1) If Roberts lost his watch during the Sports class, which one of the following must be true?
Answer: A) He lost his watch either in the morning or in the afternoon.

Reasoning: The sports class is at 10 AM or 1 PM. So, if he lost his watch during the sports class, it must have been either in the morning or in the afternoon only. Choice B is incorrect because the sports class is held on Monday and Thursday only and so, he could not have lost it in a sports class on Tuesday.

---

2) If Roberts lost his watch during the Math class, which of the following cannot be true?
Answer: A) He lost his watch in the afternoon.

Reasoning: Note that the correct answer is one that cannot be true. The math class is held only in the morning at 9 AM. So, he could not have lost his watch in the afternoon. Choice B must be true and so, it not the correct answer.

---

3) If Roberts lost his watch during the Reading class, which of the following must be true?
Answer: A) He lost his watch either on Monday or on Wednesday

Reasoning: The reading class is on Monday and Wednesday. So, if Roberts lost is watch during the Reading class, he must have lost it either on Monday or on Wednesday. Choice B is incorrect because he could have lost it at the 8 AM Reading class, which is before noon.

Answers

© Gift Of Logic, Inc * Copying prohibited

1. GROUPING AND SUMMARIZING

The list below shows the contents of a bag.

Contents	Quantity
Grape	3
Carrot	2
Orange	2
Onion	4
Apple	4
Cauliflower	3

Fill in the grid below that shows vegetables and fruits grouped together.

	Vegetables	Quantity	Fruits	Quantity
	Carrot	2	Grape	3
	Onion	4	Orange	2
	Cauliflower	3	Apple	4
Total	Total vegetables	9	Total fruits	9

There are more number of vegetables than there are fruits.
    Answer: False. There are 9 vegetables and 9 fruits.

Answers

1 VENN DIAGRAM

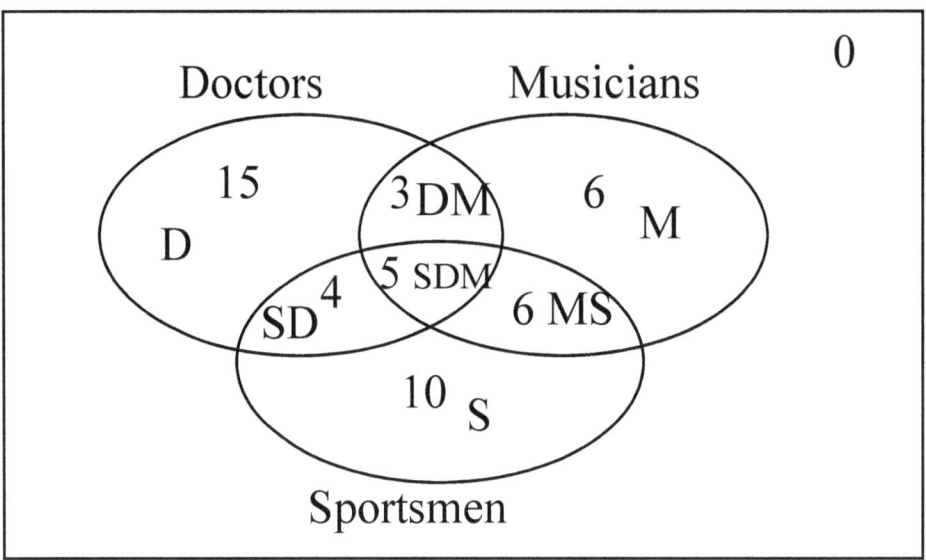

To avoid confusion, write the first letter of each group in the different areas of the Venn diagram. D,M and S represent doctors only, musicians only and sportsmen only. DM represents those that are doctors and musicians; MS represents musicians and sportsmen; SD represents sportsmen and doctors. Note that since everyone in the room belongs to one of the groups, we can place a 0 outside these groups.

1) How many doctors are not musicians or sportsmen?
   D=15 doctors
2) How many are sportsmen as well as doctors?
   SD + SDM= 4 +5 = 9
3) How many sportsmen are neither doctors nor musicians?
   S=10 sportsmen
4) How many people are there in the room?   Add all the people = 49
5) How many are either doctors or musicians?
   D+ SD+ DM+ M+ MS+ SDM = 15+4+3+6+6+5=39
6) How many are sportsmen and doctors only? SD = 4

Answers
© Gift Of Logic, Inc * Copying prohibited

2 VENN DIAGRAM

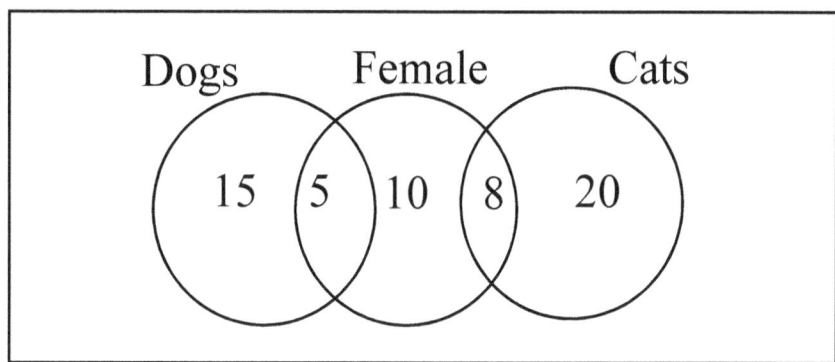

All the animals in a pet care facility can be represented by the Venn diagram shown above. Answer the questions based on the Venn diagram.

1) How many dogs are female?
   5 dogs.

2) There are no animals other than cats and dogs.
   B) False - there are 10 animals that are neither cats nor dogs.

3) How many animals are female?
   5+10+8=23 animals.

4) How many male dogs are there in the pet care facility?
   15 dogs.

5) How many cats are there in the pet care facility?
   8+20 = 28 cats.

6) How many animals are there in the pet care facility?
   15+5+10+8+20=58 animals

Answers

© Gift Of Logic, Inc * Copying prohibited

3 VENN DIAGRAM

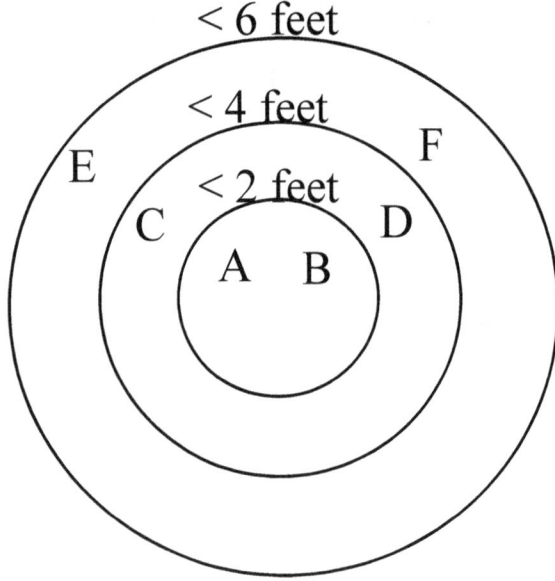

Place the members inside the Venn Diagram.

1) Who are less than four feet tall?
 These are the ones that are inside the 4 feet and 2 feet circles -A,B,C,D

2) Who are more than two feet, but less than six feet tall?
    The ones that are outside the 2 feet circle - C,D,E,F

3) Who are not less than four feet tall?
The ones outside the 4 feet circle-E,F. Note carefully the "not" in the question.

4) Who are less than two feet tall or more than 4 feet, but less than 6 feet tall?   less than 2 feet - A,B ; more than 4 feet, but less than 6 feet - E,F
    So, less than 2 feet or more than 4 feet, but less than 6 feet = A,B,E,F

Note carefully the "or" in the question.

Answers
© Gift Of Logic, Inc * Copying prohibited

1 GRAPH LOGIC

The sales data of a car manufacturing company is shown in the bar chart above.

1) The one year decline in sales from 1966 to 1967 was less than the one year increase in sales in from 1967 to 1968.
Answer: B) False. The one year decline in sales from 1967 to 1968 is 40 cars. The one year increase in sales from 1967 to 1968 is 10 cars. There was more decline than increase in sales.

2) From 1966 to 1969, there were more growth periods than periods of decline in car sales.
  Answer: A) True. There were 2 growth periods, 1967-1968 and 1968-1969 whereas there was only one decline period, 1966-1967.

2 GRAPH LOGIC

A small book store recorded sales of fictional and non-fictional books as shown above for the months of May, June and July.

1) There was a 50% decrease in non-fiction book sales from May to June and June to July.
Answer: B) False. There was a 50% decrease in non-fiction book sales from June to July (2 to 1), but not from May to June (3 to 2).

2) There was a 50% increase in fiction book sales from June to July.
Answer: A) True. In June the sales was 2 books and in July the sales was 3 books, a 50% increase from June.

3 GRAPH LOGIC

A nursery sold flowers in bunches of 50 and 100 and displayed the following symbols.

1) Garden-2 has more flowers than garden-1.
Answer: B) False. Garden-2 has 200 flowers, but garden-1 has 250 flowers.

2) Garden-1 has more 100-flower bunches than garden-2.
Answer: A) True. Garden-1 has two 100-flower bunches, but garden-2 has only one 100-flower bunch.

4 GRAPH LOGIC

The rainfall recording for an area are shown in the chart above.

1) The number of times it rained 2" or below in October is the same as the number of times it rained 3" or below in November.

Answer: A) True. It rained 2 inches or below 7 times (4+3) in October and 3 inches or below 7 times (3+4) in November.

2) 2) It rained 3" more number of times than it rained 1".
Answer: B) False. It rained 3 inches 4 times (1+0+3) whereas it rained 1 inch 7 times (4+3+0).

Answers
© Gift Of Logic, Inc * Copying prohibited

5 GRAPH LOGIC

The bar graph show the rainfall and evaporation amounts..
Note that water retention is the difference between rainfall and evaporation.

1) More water evaporated in 2006 than in 2007. Answer: A) True.
20 mm of water evaporated in 2006 against 15 mm in 2007.

2) More water was retained in 2007 than in 2006.
Answer: B) False. In 2006, 30-20=10 mm of rainfall was retained. In 2007, 20-15=5 mm of rainfall was retained.

6 GRAPH LOGIC

The trajectory of two airplanes, A and B..
1) Airplane-A climbed to 25000 ft faster than airplane-B.
Answer: B) False. Airplane-A took 1 hour to climb 25000 feet, but airplane-B took only thirty minutes to do so.

2) Airplane-B took more time to descend 25000 feet than airplane-A.
Answer: B) False. Airplane-A started its descent from 25000 feet at 2.5 hours after takeoff and landed at 3 hours after takeoff- meaning that it took 0.5 hours for the descent. Airplane-B started its descent at 3.5 hours and landed at 4 hours, also taking 0.5 hours for its descent. So, the two airplanes took the same time to descend.

3) Both airplanes cruised for the same duration.
Answer: B) False. Airplane A-cruised for 1.5 hours whereas airplane-B cruised for 3 hours. Cruising is the phase after ascent.

Answers

© Gift Of Logic, Inc * Copying prohibited

7 GRAPH LOGIC

Temperature in city-A rose at a constant rate from 60 degrees at 6 AM to 70 degrees at 4 PM.

1) If the temperature in city-B is 55 degrees at 6 AM and increased at the same rate as city-A, then city-B will be warmer than city-A at 4 PM.

Answer: B) False. At a rate of 1 degree for every one hour, city-B will see its temperature increase from 55 degrees at 6 AM to 65 degrees at 4 PM. This is not warmer than city-A which will have 70 degrees at 4 PM.

2) If the temperature in city-B is 55 degrees at 6 AM and increased at twice the rate as city-A, then both cities will have the same temperature at 11 AM.

Answer: A) True. At 11 AM, temperature in city-A will be 65 degrees. At the rate of increase of 2 degrees per hour, temperature in city B also will be 65 degrees at 11 AM.

Answers

NUMBER LOGIC

Figure out the logic in the sequence and find the missing number.

1
1.25   1.50   ?   2   Answer: 1.75 - Numbers increase by 0.25.

2
6   2   4/3   ?   Answer: 4/3 - numbers get multiplied by 1/3, 2/3, 3/3 to get the next number.

3
9   ?   8   7.5   Answer: 8.5 - numbers decrease by 0.5

4
1   1/2   1/4   ?   Answer: 1/8 - numbers get halved to get the next number

5
8   -8   16   ?   Answer: -48 - numbers get multiplied by -1, -2, -3 to get the next number

6
10   -10   20   ?   Answer: -60 - number get multiplied by -1, -2, -3 to get the next number

7
1   2   1   3   1   4   ?   Answer: 1 - multiply by 2, divide by 2, multiply by 3, divide by 3 etc.

Answers
© Gift Of Logic, Inc * Copying prohibited

NUMBER LOGIC

Figure out the logic in the numbers and find the missing number.

8 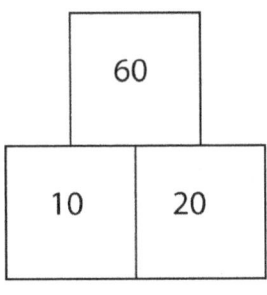	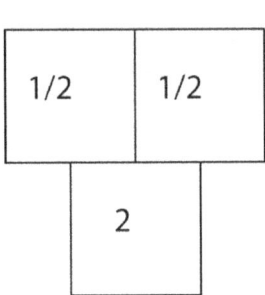	LHS = figure on left hand side. RHS= figure on right hand side. On LHS, (10+20)*3=60. On RHS, (5+8)*2=26
9 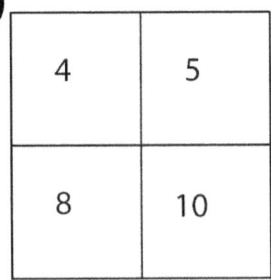 (bottom shows 2)	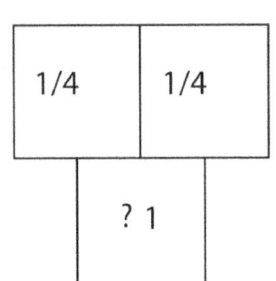	On LHS, (1/2+1/2) * 2 =2. On RHS, (1/4+1/4)*2=1
10		On LHS, 4*2=8 and 5*2=10. On RHS 3*2=6 and 9*2=18.
11 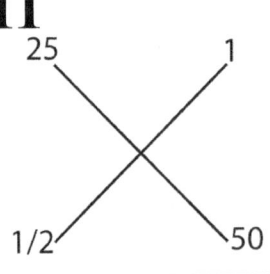	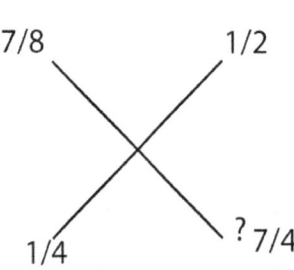	On LHS, 25*2=50, 1*1/2 = 1/2. On RHS, 7/8*2=7/4 and 1/2*1/2=1/4.
12 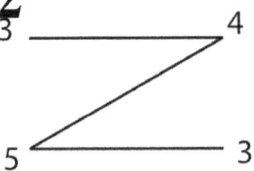		On LHS, 4+1=5, 3*1=3. On RHS, 7+1=8, 6*1=6.

Answers

© Gift Of Logic, Inc * Copying prohibited

NUMBER LOGIC

Figure out the logic in the numbers and find the missing number.

13 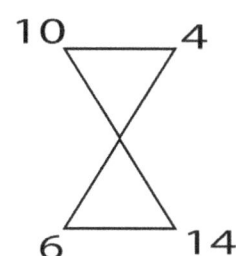	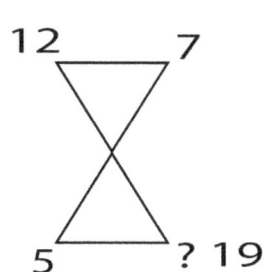	LHS = figure on left hand side. RHS= figure on right hand side. On LHS, 10-4=6, 10+4=14. On RHS, 12-7=5 and 12+7=19.
14 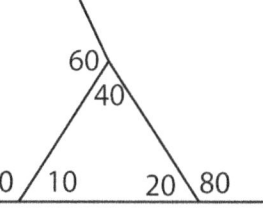	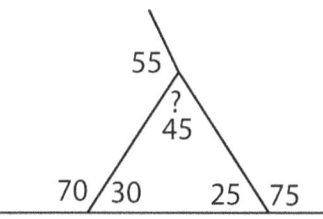	On LHS, all the pairs of numbers add up to 100. On RHS, 55+ 45=100.
15 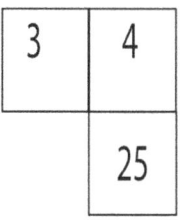	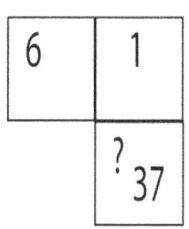	On LHS, 3*3+ 4*4=25, On RHS, 6*6 + 1*1=37.
16 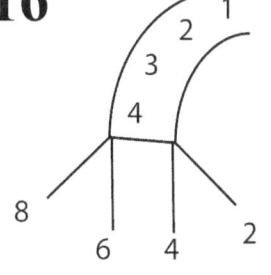	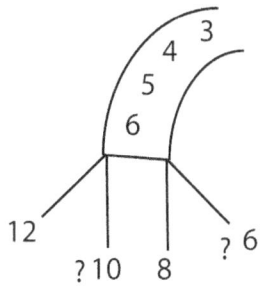	On LHS, each number in the pipe gets multiplied by 2 and falls down. On RHS, 3*2=6 and 5*2=10.

Answers

LETTER LOGIC

Figure out the logic in the sequence and find the missing letter or number.

1   A21   B63   ?   Answer: C189 - alphabets increase in sequence and numbers are multiplied by three

---

2   3C3   4D4   ?   Answer: 5E5 - alphabets and numbers increase in sequence.

---

3   26-Z   1-A   25-Y   ?   Answer: 2-B

One letter from the end and one letter from the beginning of the alphabetic sequence appears alternately, along with their numbers.

---

4   P^Q Q^R ?   Answer: R^S - The alphabets appear in sequence, with the second one being superscripted on the first one.

---

5   T/V   V/X   X/?   Answer: Z - alphabet below is two letters after the alphabet above. After X, skip one and get Z.

---

6   BCD   24   EFG   ?   Answer: 210 - B,C,D are the 2nd, 3rd and 4th letters. 2*3*4=24. E,F,G are the 5th, 6th and 7th letters. 5*6*7=210.

---

7   BDF   12   CEG   15   MOQ   ?   Answer: 45 - B,D,F are the 2nd, 4th and 6th letters. 2+4+6=12. C,E,G are the 3rd, 5th and 7th letters. 3+5+7=15. M,O,Q corresponds to 13th, 15th, and 17th letters. 13+15+17=45.

Answers

LETTER LOGIC

Figure out the letter or number indicated by the question mark ?

8 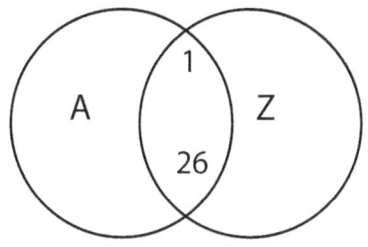 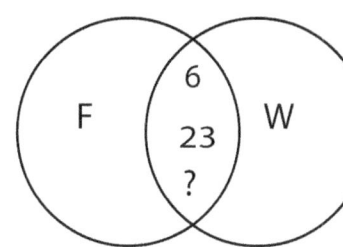 On LHS, A corresponds to 1 and Z corresponds to 26. On RHS, F corresponds to 6 and W corresponds to 23.

9 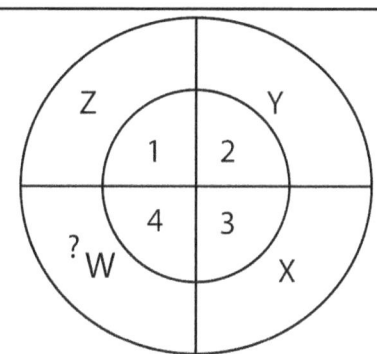 On LHS, Z is the 1st letter from the end, Y is the second letter from the end, X is 3rd and W is the 4th letter from the end.

10

L	M	N
N	O	? P - each letter below is obtained by skipping one letter from the one above.

11

A	E	I
C	G	? K - each column has letters that skip one letter
E	I	M

12
If A=1, B=2 and so on, then JACK is
    Answer: B) 101311 10 = J, 1=A, 3=C, K=11
If A=1, B=2 and so on, then LAD is
    Answer C) 1214   L=12, A=1, D=4

Answers
© Gift Of Logic, Inc * Copying prohibited

LETTER LOGIC

13 Figure out the logic in the sequence and find the missing letter or number.

1, 2, C   5, 6, G   8, 9, ?   Answer: J - 1,2,C -- the third entry in each set is the alphabet corresponding to the number. 3 should have been the third number in the first set, but it is replaced with the corresponding letter C. Similarly, 5,6,7 in the second set is shown as 5,6,G, G being the 7th letter. Similarly, 8,9,10 and the 10th letter is J.

14

If G is A and D=M then GOOD = Answer: A) AOOM. Substitute A for G and M for D.

---

15 Aliens visiting earth..

1) What do the aliens mean when they say TRAMS?
Answer: C) SMART. They mean the reverse of what they say. RAC reversed in CAR and ESOUM reversed is MOUSE. So, TRAMS reversed in SMART.

2) What do the aliens mean when they say CIGOL?
Answer: B) LOGIC. This is the English word that you get when you reverse the alien word CIGOL.

3) What do the aliens say when they want to refer to a BOY?
Answer: B) YOB. This is the reverse of BOY.

Answers

© Gift Of Logic, Inc * Copying prohibited

1 SUDOKU

Solve the following Sudoku. A correctly solved Sudoku has numbers 1-9 appearing only once in each row, each column and each 3x3 grid. You gain valuable positioning skills by solving these sudokus.

5	7	8	3	2	1	4	6	9
9	3	4	6	5	7	1	2	8
6	1	2	4	9	8	5	7	3
8	9	5	7	4	3	2	1	6
3	4	7	2	1	6	8	9	5
1	2	6	9	8	5	7	3	4
2	5	1	8	3	9	6	4	7
7	8	9	1	6	4	3	5	2
4	6	3	5	7	2	9	8	1

Answers

© Gift Of Logic, Inc * Copying prohibited

2 SUDOKU

Solve the following Sudoku. A correctly solved Sudoku has numbers 1-9 appearing only once in each row, each column and each 3x3 grid. You gain valuable positioning skills by solving these sudokus.

1	9	7	8	5	6	4	3	2
5	6	8	2	3	4	7	9	1
2	4	3	7	9	1	8	5	6
9	3	2	6	1	8	5	4	7
7	5	4	3	2	9	6	1	8
8	1	6	4	7	5	9	2	3
6	8	9	1	4	3	2	7	5
4	7	1	5	6	2	3	8	9
3	2	5	9	8	7	1	6	4

Answers

© Gift Of Logic, Inc * Copying prohibited

3 SUDOKU

Solve the following Sudoku. A correctly solved Sudoku has numbers 1-9 appearing only once in each row, each column and each 3x3 grid. You gain valuable positioning skills by solving these sudokus.

3	1	2	8	5	4	6	9	7
7	5	9	6	2	1	8	4	3
6	8	4	9	7	3	1	2	5
5	7	6	4	3	8	9	1	2
9	3	1	5	6	2	4	7	8
2	4	8	7	1	9	5	3	6
8	6	3	1	4	7	2	5	9
4	2	5	3	9	6	7	8	1
1	9	7	2	8	5	3	6	4

Answers

4 SUDOKU

Solve the following Sudoku. A correctly solved Sudoku has numbers 1-9 appearing only once in each row, each column and each 3x3 grid. You gain valuable positioning skills by solving these sudokus.

9	8	4	7	6	5	3	1	2
6	5	2	9	3	1	8	4	7
7	1	3	8	4	2	9	5	6
8	3	6	1	5	4	2	7	9
4	9	5	2	7	3	1	6	8
2	7	1	6	9	8	4	3	5
1	6	8	3	2	7	5	9	4
3	4	9	5	8	6	7	2	1
5	2	7	4	1	9	6	8	3

Answers

© Gift Of Logic, Inc * Copying prohibited

5

SUDOKU

Solve the following Sudoku. A correctly solved Sudoku has numbers 1-9 appearing only once in each row, each column and each 3x3 grid. You gain valuable positioning skills by solving these sudokus.

5	7	8	6	4	3	1	2	9
1	2	6	8	7	9	4	5	3
4	9	3	2	5	1	7	6	8
6	8	9	1	3	5	2	7	4
3	1	5	4	2	7	9	8	6
2	4	7	9	8	6	5	3	1
9	6	2	7	1	8	3	4	5
8	5	4	3	9	2	6	1	7
7	3	1	5	6	4	8	9	2

Answers

6 SUDOKU

Solve the following Sudoku. A correctly solved Sudoku has numbers 1-9 appearing only once in each row, each column and each 3x3 grid. You gain valuable positioning skills by solving these sudokus.

6	3	1	7	5	2	4	8	9
5	4	9	3	8	1	7	6	2
2	8	7	4	9	6	1	3	5
9	5	4	8	6	7	3	2	1
3	2	8	5	1	9	6	7	4
7	1	6	2	3	4	9	5	8
4	7	3	1	2	8	5	9	6
1	6	2	9	7	5	8	4	3
8	9	5	6	4	3	2	1	7

Answers
© Gift Of Logic, Inc * Copying prohibited

PICTURE SEQUENCE

Figure out the logic in the picture sequence, and draw the next picture in the sequence.

1

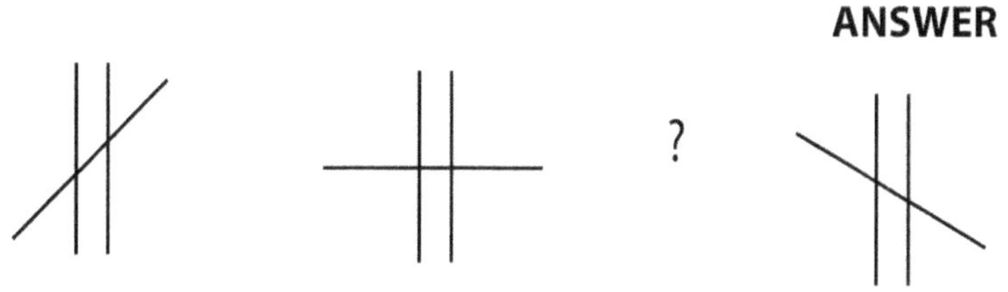

2

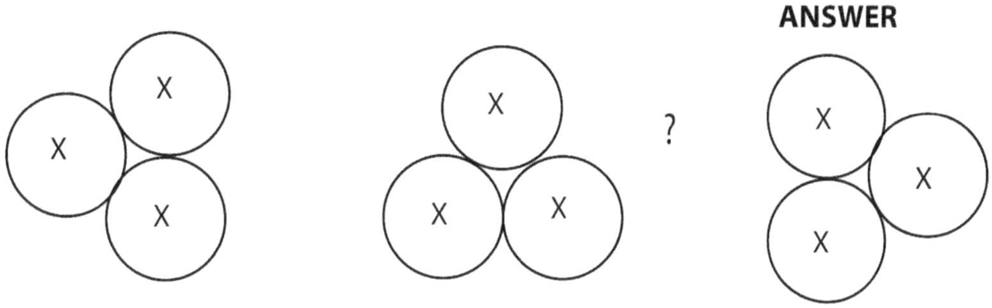

3

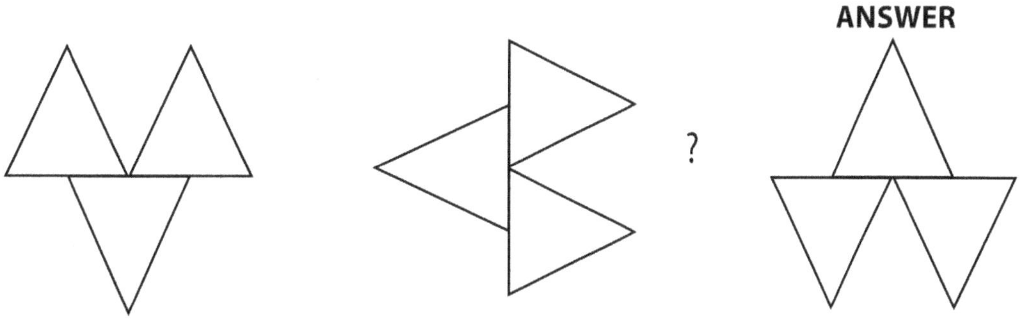

4

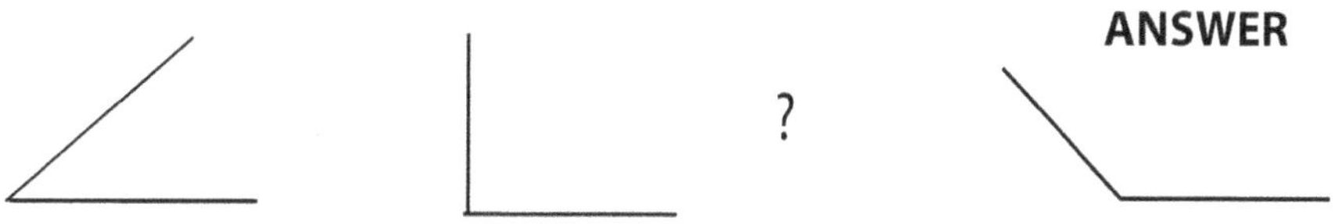

PICTURE SEQUENCE

Figure out the logic in the picture sequence, and draw the next picture in the sequence.

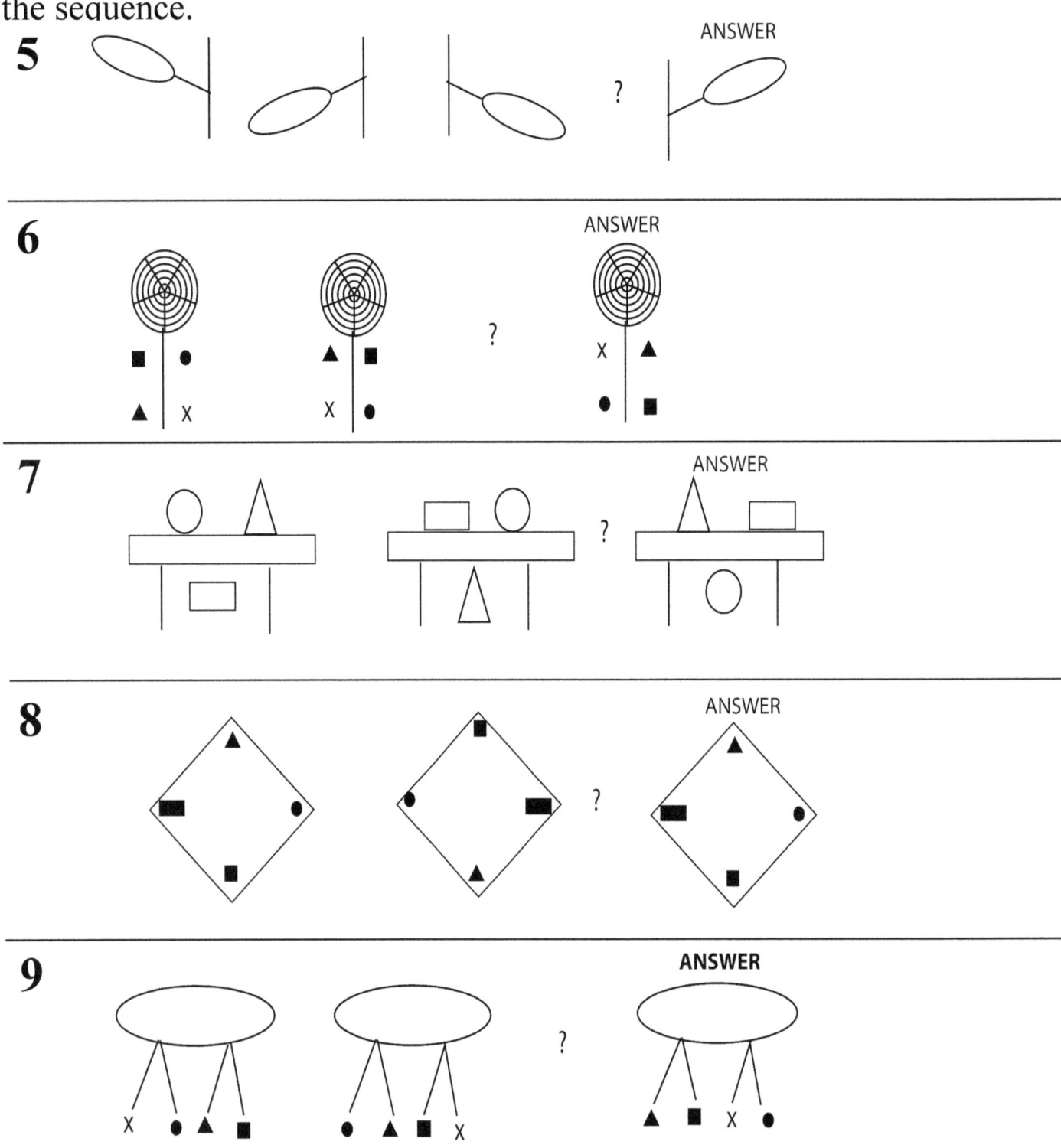

PICTURE SEQUENCE

Figure out the logic in the picture sequence, and draw the next picture in the sequence.

10

  ? 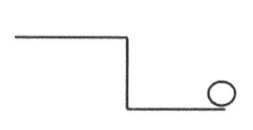 **ANSWER**

11

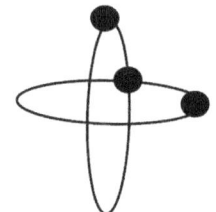 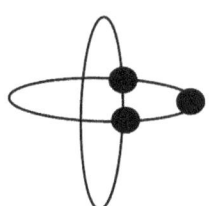 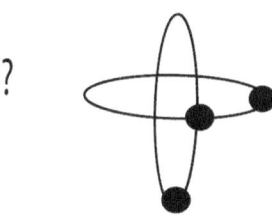 ? **ANSWER**

12

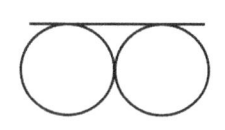 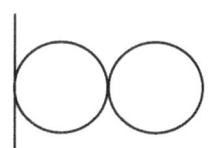 ? 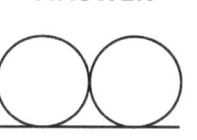 **ANSWER**

13

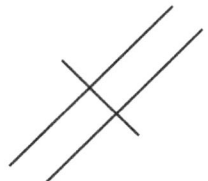 ? **ANSWER**

14

? **ANSWER**

Answers 127

© Gift Of Logic, Inc * Copying prohibited

PICTURE ANALOGY

Figure out the logic in the picture analogy, and draw the correct picture that will complete the analogy.

1

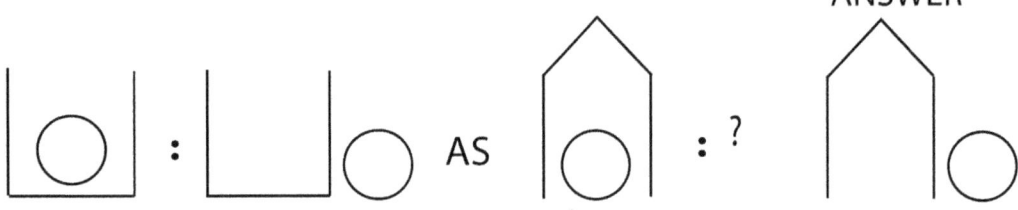

2

3

4

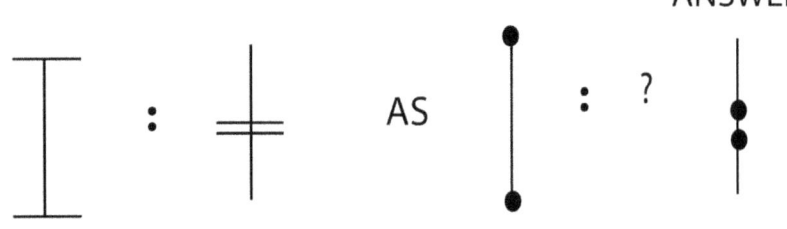

5

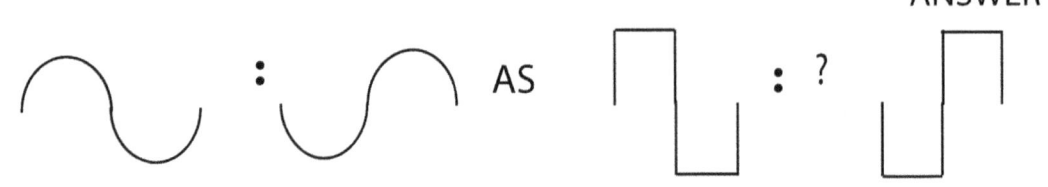

PICTURE ANALOGY

Figure out the logic in the picture analogy, and circle the correct picture that will complete the analogy.

6

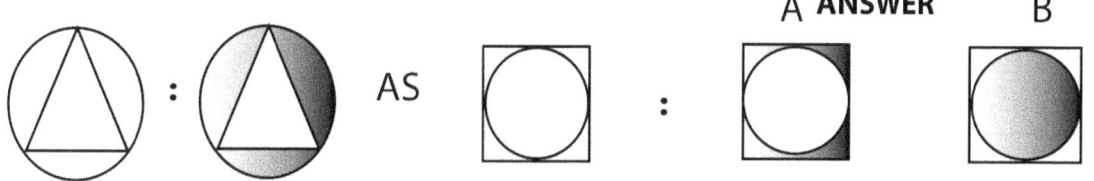

7

8

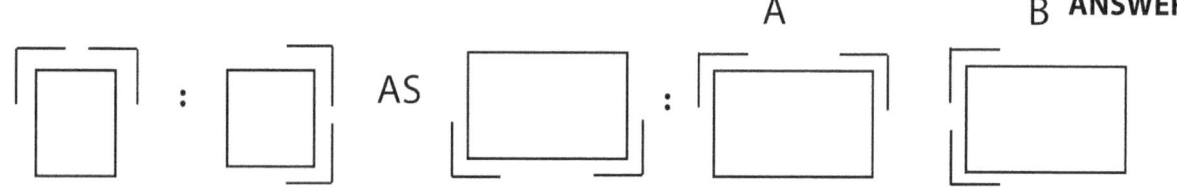

9

PICTURE ANALOGY

Figure out the logic in the picture analogy, and circle the correct picture that will complete the analogy.

10 : 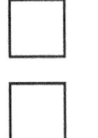 AS 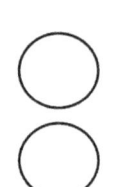 :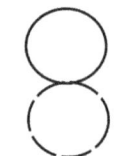

A   B   **C ANSWER**

11  :  AS 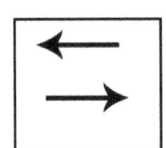 :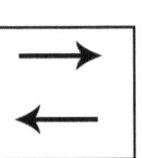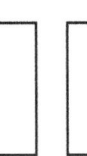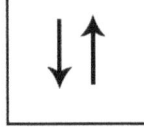

A   B   **C ANSWER**

12  AS 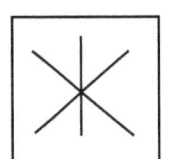 :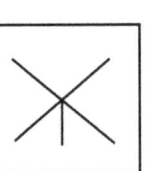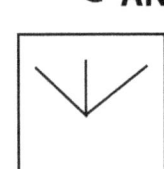

A   B   **C ANSWER**

13 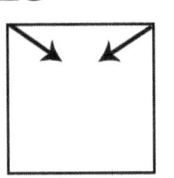 : 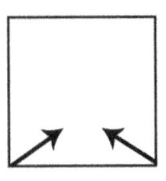 AS 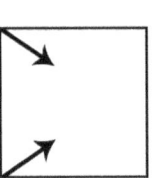 :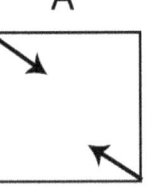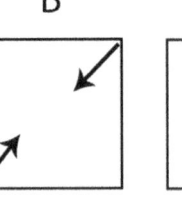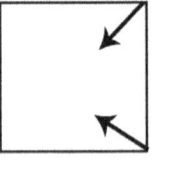

A   B   **C ANSWER**

14  AS  :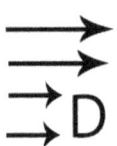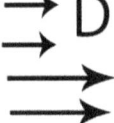

A   **B ANSWER**

Answers

© Gift Of Logic, Inc * Copying prohibited

ODD PICTURE

Figure out the logic in the pictures and identify the odd picture.

1 A   B   C **ANSWER**

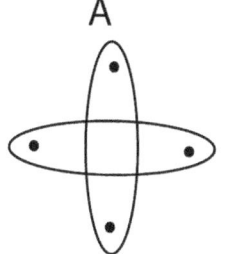    dots are inside in C

2 A   B   C **ANSWER**

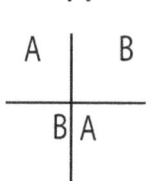 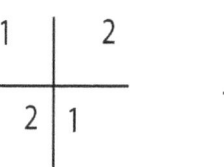 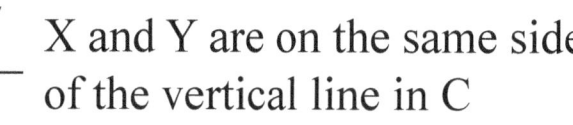 X and Y are on the same side of the vertical line in C

3 A   B **ANSWER**   C

 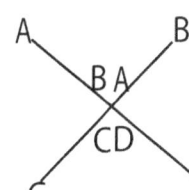   symbols $ and X are not reversed in B

4 A   B **ANSWER**   C

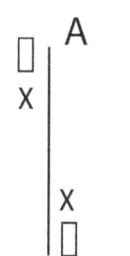 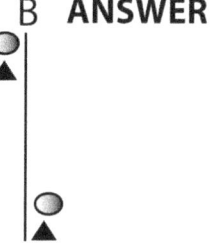   symbols are not reversed in B

5 A   B   C **ANSWER**

 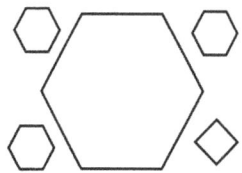   all the outside shapes are not the same in C

Answers
© Gift Of Logic, Inc * Copying prohibited

ODD PICTURE

Figure out the logic in the pictures and identify the odd picture.

6 A   B   C **ANSWER**

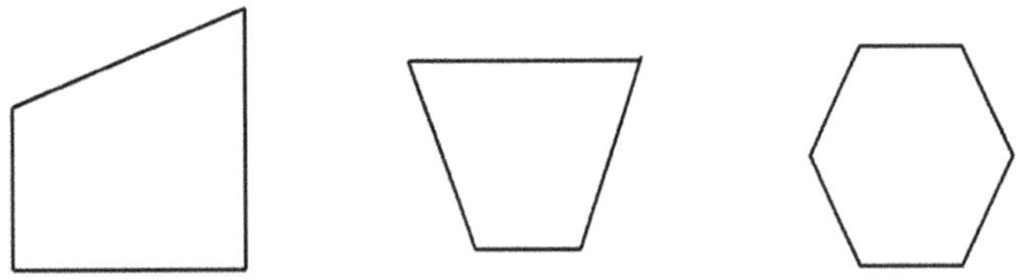

five sides in C

7 A   B   C **ANSWER**

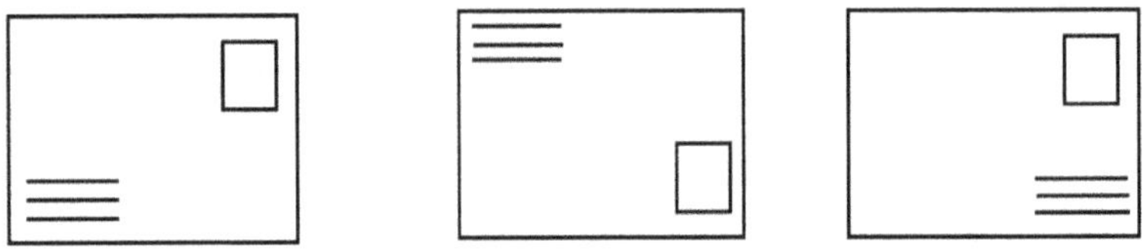

square and lines are on same side in C

8 A   B **ANSWER**   C

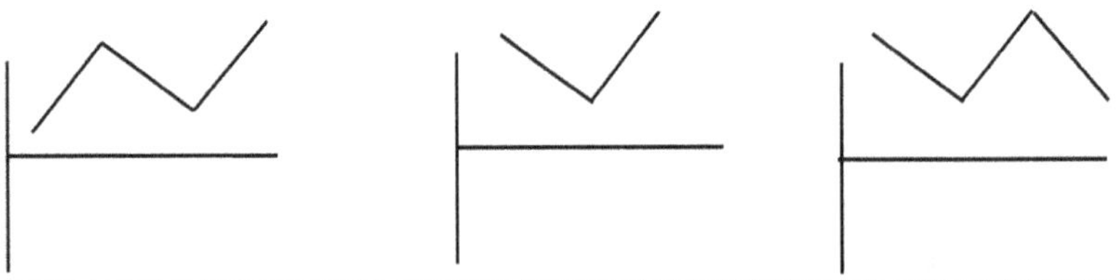

two legs in graph in B

9 A   B   C **ANSWER**

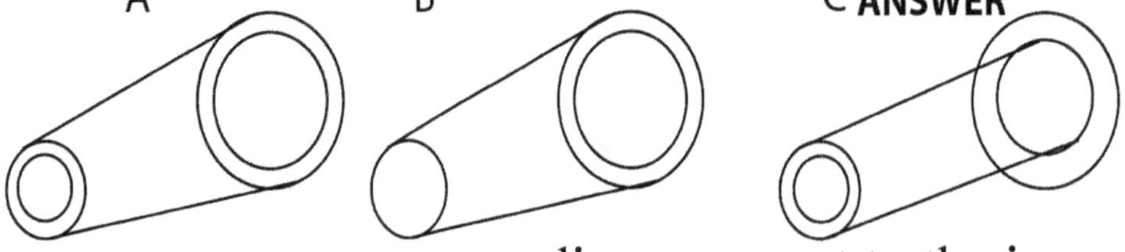

lines connect to the inner circle in C

10 A   B **ANSWER**   C

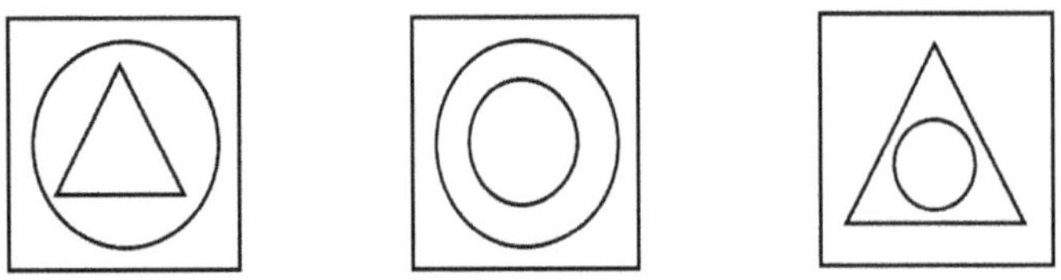

identical shapes inside the square in B

Answers

© Gift Of Logic, Inc * Copying prohibited

132

PICTURE DIFFERENCE

Mark the differences in the set of pictures shown with arrows.

1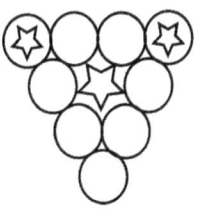

2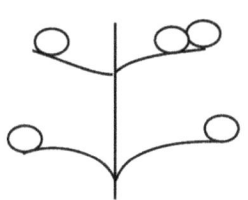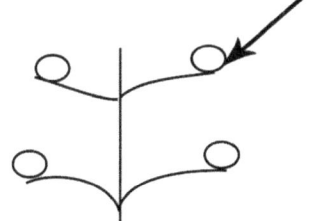

3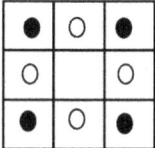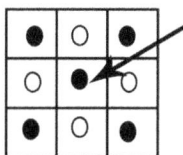

4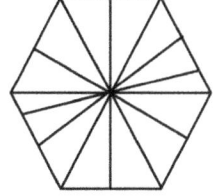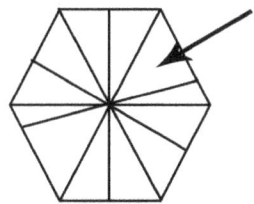

5

Answers

PICTURE DIFFERENCE

Mark the differences in the set of pictures shown with arrows.

6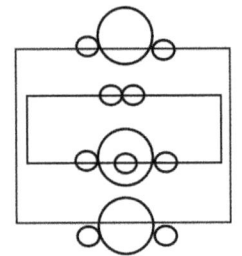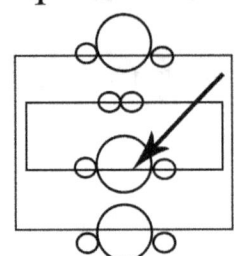

7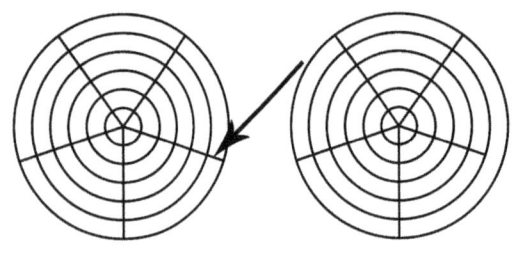

8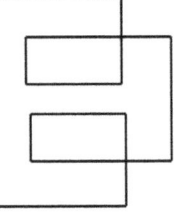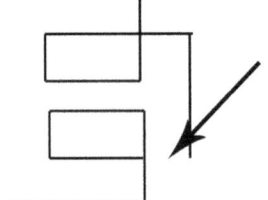

9

10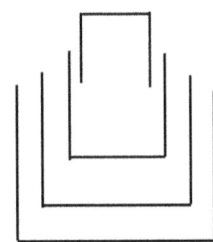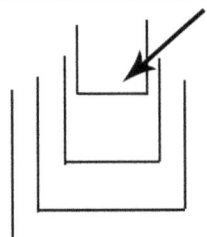

Answers

PICTURE DIFFERENCE

Mark the differences in the set of pictures shown with arrows.

11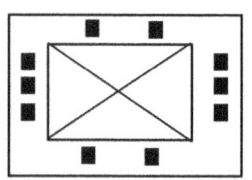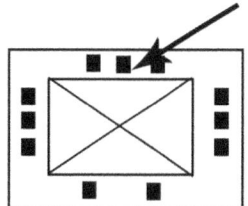

12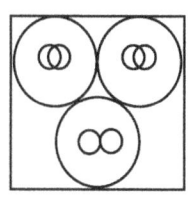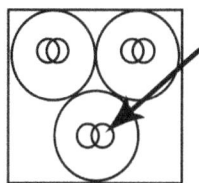

13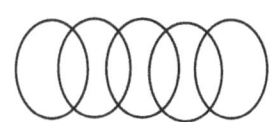

14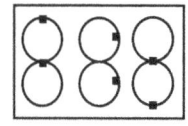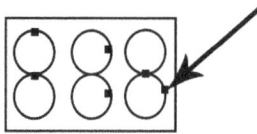

15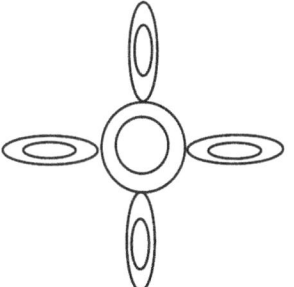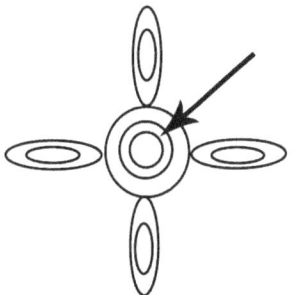

PATTERN MATCHING

Find the logical pattern in the pictures on the left and identify the picture on the right that will fit in the space marked with ? to complete the pattern.

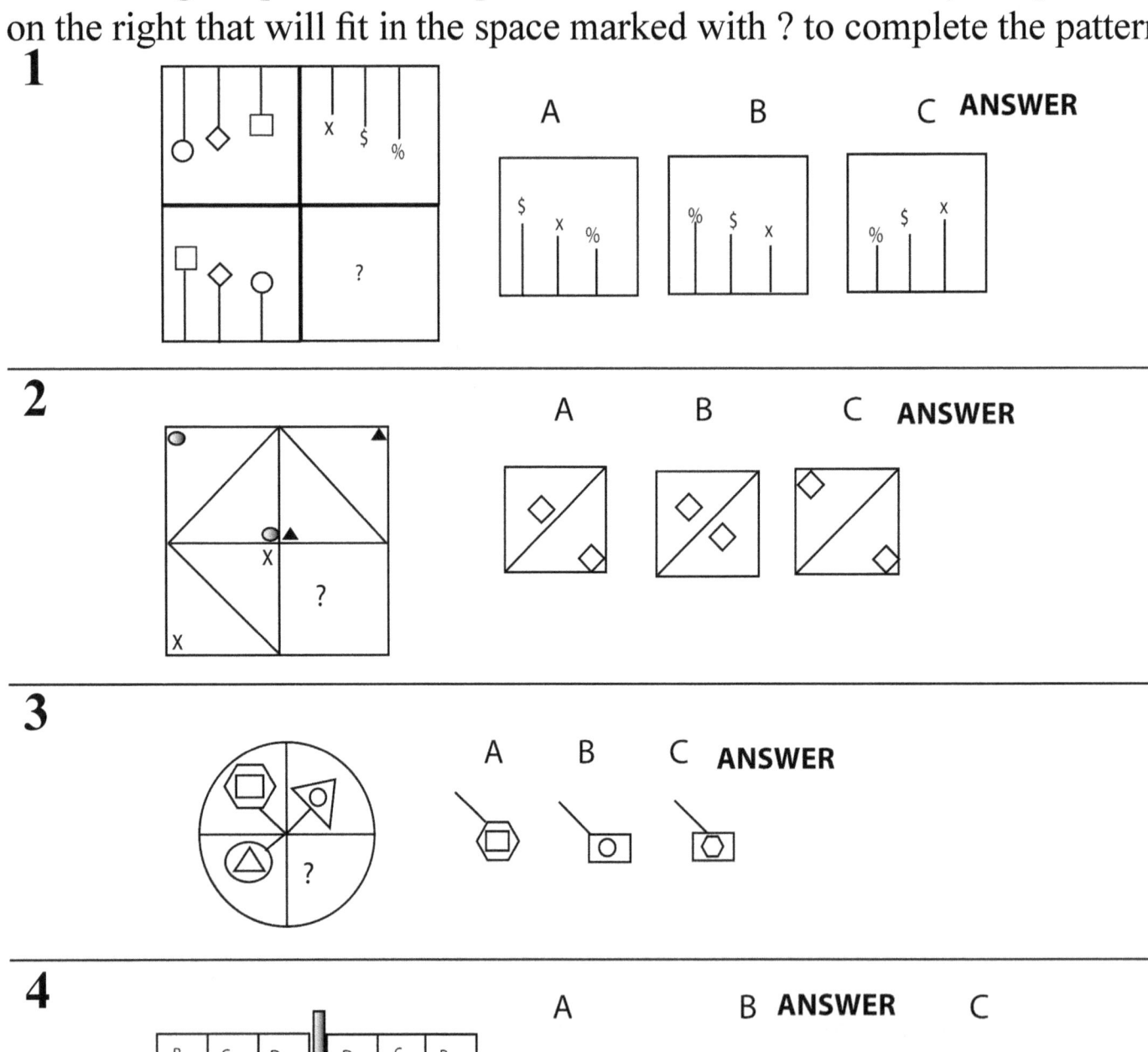

NOTES

NOTES

NOTES

www.ingramcontent.com/pod-product-compliance
Lightning Source LLC
Chambersburg PA
CBHW080256180526
45167CB00006B/2545